CONTENTS

LIST OF ILLUSTRATIONS

LIST OF TABLES

ACKNOWLEDGMENTS

The basic research upon which this publication is based was made possible by a grant from the National Science Foundation, Number SOC76-82632. Facilities were provided by the Department of Geography at the University of Chicago.

The authors are:

Norton Ginsburg, Professor of Geography, the University of Chicago.

James Osborn, Assistant Director for Project Development and Regional Implementation, Regional Economic Development Office for West Africa, AID, Abidjan.

Grant Blank, Doctoral Candidate in Sociology, the University of Chicago.

A number of individuals made substantial contributions to the research effort. Among these, special appreciation is due to: Marilyn Dorn, the Department of State; Daniel Dzurek, the Department of State; Richard L. Edmonds, the School of Oriental and African Studies, the University of London; and Richard Forstall, the Bureau of the Census.

The views expressed here are those of the authors and should not be attributed to the National Science Foundation, AID, the Bureau of the Census, or the Department of State. Errors of fact and interpretation are, of course, the authors' own.

CHAPTER I

SOME EMPIRICALLY BASED GENERALIZATIONS

Introduction

The search for a better way to describe--and to map--the distribution of wealth and poverty in the world has challenged investigators for decades. The patterns that emerge from any global analysis are characterized by striking inequalities. Since the wealth of a nation is intimately related to vital issues such as health, longevity, and the range of choices and comforts available to people, the inequalities continue to attract attention. Social scientists have devoted much effort to describing these inequalities, to explaining why they have occurred, and to measuring them.

How wealth is measured plays a crucial role in formulating conclusions that illuminate pattern and process. Comparative analysis of conditions in a variety of countries plays an essential role in attempting to understand these inequalities and the processes that underlie them. Through such comparisons, the unique features of individual nation-states may be recognized, even as patterns of uniformity are identified. Despite the widespread use of Gross National Product per capita as a primary measure of relative prosperity, it is clear that prosperity is both complex and multidimensional. It can be addressed only by a variety of techniques and a number of indicators. This premise provides the rational for the following inquiry into global patterns of wealth and poverty.

As one of a number of approaches to this problem, in 1961, the *Atlas of Economic Development* ranked and grouped countries according to arrays of characteristics that theory and common sense suggested were important in discriminating among countries which might, on the face of it, fall within the same categories of wealth and poverty as measured by GNP per capita alone.[1] In that volume also an attempt

---

1. Norton Ginsburg, *An Atlas of Economic Development* (Chicago: University of Chicago Press, 1961), with Preface by B.F. Hoselitz and Statistical Appendix

was made to identify underlying relationships among country
characteristics on the one hand and, on the other, the ways
countries clustered statistically across these underlying
dimensions. Both attempts were successful, but their explanatory
and interpretive powers were limited. For one thing, time series
data were few. For another, the analytical techniques employed were
experimental, if effective for their time.

This study applies rather more sophisticated techniques to the
problem and thereby seeks to further illuminate the nature of
development at national, regional, and global scales. In the course
of the the analysis some light is cast also on the relationships
among variables commonly used to assess developmental status, and
maps are presented which help introduce the reader to a more subtle,
and possibly more significant, world economic geography than that to
which they had been accustomed.

Specifically, the objectives of the study are to identify
again some basic patterns of wealth and poverty by country on a
global basis, to examine the evidence for regional developmental
structures and problems, to measure the distinctiveness of countries
with regard to their socio-economic structures, and to devise means
for comparing countries in terms that will complement conventional
measures of differences.

To these ends, data for 163 variables were compiled and
analyzed for up to 185 country units.[2] Some 38 of these refer
specifically to international development assistance as a specific
aspect of the problem and will not concern us further at this point.
Of the remaining 125, 83 variables, or indicators of developmental
characteristics, were mapped and carefully studied so as to identify
overlap and redundancies. Consequently, this number was reduced to
66 (tables 1 and 2), and these were subjected to a principal
components analysis covering 143 countries (see appendix A). As a
result, a number of "factors" were identified, mapped, and analyzed.
Several "World Standard" indices also were devised for measuring the
"distance" between the poorer countries and the wealthier. The
results suggest significant differences in developmental structures
and prospects both at regional and individual country levels.

---

by B.J.L. Berry. The atlas also appeared as Research Paper no. 68, University of
Chicago, Department of Geography, 1960.

2. For lists of countries and variables, see appendices A and B. Following
usage in the 1961 *Atlas of Economic Development,* p. 15, countries are grouped into
10 realms: Australasia, Anglo-America, Latin America, Western Europe, Eastern
Europe, Near East, Sub-Saharan Africa, South Asia, Southeast Asia, and East Asia.

TABLE 1

COMPLEXES OF VARIABLES

I. The Product and Productivity Complex

   A. National Product

      1. Gross National Product
      2. Gross National Product per Capita
      3. Gross National Product per Capita Growth Rate, 1960-1975
      4. Gross National Product per Capita Growth Rate, 1970-1975
      5. Manufacturing as a Percent of Gross Domestic Product
      6. Agricultural as a Percent of Gross Domestic Product

   B. Investment and Productivity

      1. Gross Domestic Investment as a Percent of Gross Domestic Product
      2. Gross Domestic Investment Growth Rate, 1965-1973
      3. Agricultural Production Growth Rate, 1960-1970
      4. Agricultural Production Growth Rate, 1970-1974
      5. Wheat Yields
      6. Paddy Yields
      7. Consumer Price Index Change, 1970-1975

II. The Technology and Consumption Complex

   A. Technology and Energy

      1. Gross Energy Consumption
      2. Gross Energy Consumption per Capita
      3. Commercial Energy as a Percent of Gross Energy Consumption
      4. Commercial Energy Consumption per Capita
      5. Commercial Energy Consumption per Capita Growth Rate, 1965-1975
      6. Commercial Energy Consumption Growth Rate, 1965-1975
      7. Electricity Generation per Capita

   B. Technology and Industrial Products

      1. Steel Consumption per Capita
      2. Motor vehicles per Capita
      3. Tractors per Capita Agricultural Population
      4. Tractors per Unit of Arable Land
      5. Fertilizer Consumption per Capita Agricultural Population
      6. Fertilizer Consumption per Unit of Arable Land

III. The Population Complex: Structure, Organization, and Quality

   A. Demographic Aggregates

      1. Total Population, 1975
      2. Population Growth Rate, 1965-1975
      3. Population Density

   B. Internal Structure

      1. Youthfulness (Percent of Population Age 14 or Less)
      2. Percent of Economically Active Population in Agriculture
      3. Agricultural Population as a Percent of Total Population
      4. Agricultural Population Growth Rate, 1965-1975
      5. Agricultural Population per Unit of Arable Land
      6. Urban Population as a Percent of Total Population: Census Definition

TABLE 1, *Continued*

    7. Urban Population as a Percent of Total Population:
       Cities of 100,000 and Greater
    8. Urban Population Growth Rate, 1964-1975
    9. Urban Primacy

C. Quality of Population

    1. Health
       a) Life Expectancy
       b) Infant Mortality
       c) Physicians and Dentists per Capita
       d) Calories per Capita per Day
    2. Education
       a) Literacy
       b) Primary School Enrollment
       c) Primary School Enrollment Growth Rate, 1960-1970
       d) Secondary School Enrollment
       e) Third-Level School Enrollment
    3. Communication
       a) Daily Newspaper Circulation
       b) Radio Ownership per Capita

IV. The Resource Endowment Complex

A. Energy

    1. Total Energy Potential
    2. Total Energy Potential per Capita
    3. Total Energy Potential per Unit Area
    4. Non-renewable Energy Potential per Capita
    5. Renewable Energy Potential per Capita
    6. Biomass Energy Potential per Capita

B. Land

    1. Arable and Cultivated Land per Capita
    2. Arable and Cultivated Land as a Percent of Total Land

V. The Accessibility and Connectivity Complex

A. Road and Water Transportation

    1. Road Length per Unit Area
    2. Road Length per Capita
    3. Motor Vehicles per Unit Road Length
    4. Inland Waterway Freight Transported

B. Rail and Air Transportation

    1. Rail Route Length per Unit Area
    2. Rail Route Length per Capita
    3. Rail Freight Ton-kilometers per Unit Route Length
    4. Rail Freight Ton-kilometers per Capita
    5. Rail and Air Passenger-kilometers per Capita

VI. The External Relations Complex

A. Total Trade

    1. Total Trade Turnover
    2. Total Trade Turnover per Capita
    3. Total Trade Turnover Growth Rate, 1965-1975

B. Outward Flows

    1. Export Dependency on Raw Materials
    2. Export Concentration
    3. Exports as a Percent of Gross National Product
    4. Exports: Percent to Industrialized Countries
    5. External Debt Ratios
    6. International Mail Outflows

To highlight the themes and concerns that underly the discussion in the succeeding chapters, the remainder of this chapter will be devoted to a broad overview. The goal is to give the reader a clearer understanding of the current state of world development at a relatively simple level, using zero-order bivariate correlations, before discussion of the more complex multivariate analyses. The overview is organized around two perspectives. Each perspective categorizes the data so as to highlight different issues. The first perspective deals with what might be called the economics of development. Here the data have been grouped into six common categories, called Complexes of Variables. The second perspective discusses the data from the point of view of eight key global and national Policy Issues.

Complexes of Variables

The literature on development has suggested a number of conceptually useful groupings of variables. In the *Atlas of Economic Development* seven groupings of variables were identified and discussed. In this case, the categories have been modified and reduced to six (table 1). They are:

1. Product and Productivity
2. Technology and Consumption
3. Population: Structure, Organization, Quality
4. Resource Endowment
5. Accessibility
6. External Relations

Within each of these complexes, the correlations among the individual variables provide useful information about the validity and utility of the complexes in assessing developmental characteristics and relationships.

The first two complexes, (1) Product and Productivity and (2) Technology and Consumption, are composed of widely accepted measures of attainment. How countries and groups of countries differ in their mixes of "success" on these indices is of obvious interest. Since these indicators tend to be intercorrelated with each other and highly correlated with gross national product, divergences among countries within these complexes highlight aberrant cases. The other four complexes are composed of indicators which may be regarded either as *conditions* for development and development policy application (e.g., resource endowment) or as *national attributes* of ambiguous yet important significance in the development process (e.g., urbanization, direction of trade). In contrast to the first two complexes these tend to be non-intercorrelated and somewhat less directly associated with GNP per capita.

The Product and Productivity Complex

For this complex, thirteen indicators (grouped into two sub-complexes: (1) National Product and (2) Investment and Productivity) have been considered. These indicators bear on a country's ability to produce, the nature of its productive capacity, and problems associated with internal finance. Within this complex of indicators, GNP/capita was compared, inter alia, to total GNP and to two GNP per capita growth measures.[3]

As a result of this analysis, it is apparent that GNP per capita in 1975 was virtually uncorrelated with GNP, with GNP per capita growth since 1960, with agricultural production growth, with gross domestic (GDI) growth, with percent of gross domestic investment in Gross Domestic Product (GDP), and with change in the consumer price index since 1970. It has a moderate negative correlation, however, with percent of GDP contributed by agriculture (-.618 for 131 countries), and a positive correlation with wheat yields (.680 for 97 countries). There are associations with GDI growth and percent of GDI in GDP (.578 for 89 countries and .520 for 99 countries, respectively) which, although significant, are not nearly so high as might have been expected. The very low general correlations (less than .300 in all but one case) of agricultural production growth variables and consumer price index change with all other indicators in this complex clearly indicate the relative independence of these phenomena, and of GNP per capita growth as well, from the patterns of current development attainment as measured by product and productivity.

Technology and Consumption Complex

Thirteen indicators of consumption of technological inputs have been considered. They were selected with the view that, although many also reflect productivity in varying degrees, it is the *availability* of, particularly, energy, but also steel, vehicles, and capital-intensive agricultural inputs, through importation or

---

3. Comparisons in this section are based on zero-order Pearson product moment correlation coefficients. Missing data were handled by pair-wise deletion. This means that for each pair of variables a different set of countries is available--the largest set which have non-missing values for both variables.

otherwise, that determines the relative success of countries in pursuing the Western model of development. Gross energy consumption per capita correlates at .998 (143 countries) with commercial energy consumption per capita, although the difference between the two measures is significant for some poor agricultural countries where the percent of gross energy consumption contributed by commercial energy is relatively small. Growth of commercial energy consumption per capita has very low positive and negative correlations with all other such indices, clustering between .186 and -.152, whereas most measures of steady-state per capita consumption of technological inputs intercorrelate rather highly (.600 to .899). Use of commercial fertilizer and tractors per unit of arable land, although highly intercorrelated with each other (at .986 for 143 countries), are not, surprisingly, significantly associated with the other per capita consumption measures.

Population: Structure, Organization, Quality Complex

This complex consists of three sub-clusters: Demographic Aggregates, Internal Structure, and Quality of Population. The first two of these are composed of twelve indicators bearing on the distribution of population within countries, especially between urban and rural areas, and concerning population growth. Four of those indicators have little or no discernible relationship with any others: total population, urban population growth, population density, and urban primacy. The others are associated with correlations of over .650 around the distinction between rural and urban populations, and there is a high degree of association among population growth, youthfulness of population, agricultural population growth, and percent of economically active population in agriculture.

The third sub-cluster includes measures of educational attainment, health, and communication levels. Apart from growth in primary school enrollments, which shows a small slightly negative association, these variables display the expected pattern: fairly uniform, moderately high, positive correlations usually between .550 and .800. The important differences between primary school enrollments and their growth reflects non-uniformity, especially among developing countries, in successfully implementing educational policy, given the location and age-structure of the target population. Higher levels of intercorrelations are found among health measures (calories per capita, life expectancy, infant mortality), higher level school enrollments, literacy, and daily newspaper circulation. Radio ownership per capita, perhaps a better measure of communication potential and spread of inexpensive technological innovation, has consistently lower though still significant correlations with the other indicators. Doctors and dentists per capita also has lower though significant correlations with the other indicators (around .6 to .7); whereas the highest correlations positive or negative are between literacy and life expectancy (.903 for 150 countries), literacy and secondary school enrollment (.843 for 126 observations), literacy and infant mortality (-.825 for 116 countries), and life expectancy with secondary school enrollment (.853 for 126 countries) and infant mortality (-.861 for 114 countries).

The pattern of correlations suggests that there is a worthwhile distinction to be made between social welfare indicators where the investments are additive per person and may accumulate around the richer and more urbanized population, e.g., indicators such as doctors, newspapers, and high-level education; and indicators of investment and consumption that reflect the more general population, such as infant mortality, life expectancy, and primary school enrollment.[4] On the other hand, the high intercorrelation of infant mortality, literacy, and life expectancy means that the PQLI (Physical Quality of Life Index) proposed by the Overseas Development Council has redundant components in a global context.[5]

These indicators have taken on a certain importance in recent years owing to the ascendance of "human needs" as one, if not the primary, focus of development policies in certain countries and development assistance agencies. To an extend this represents a shift in goals from output maximization to poverty minimization. The new goal is often seen as requiring investment that conflicts with the old--growth of productivity. However, one study indicates that even just among developing countries there is a positive association between what the author calls "human resources" indicators and growth.[6] Indeed, countries that had done well in the fulfillment of basic needs in 1960 had above average growth rates in the period 1960 to 1973. Thus, since investment in human resources may hasten growth, differentiation among indicators of development

4. See United Nations Research Institute for Social Development (UNRISD), *Contents and Measurement of Socio-Economic Development: An Empirical Enquiry* (Geneva, 1970), p. 5.

5. See Overseas Development Council, *The United States and World Development, Agenda 1977* (New York: Praeger, 1977), pp. 197 ff.

6. See Norman Hicks, *Economic Growth and Human Resources,* World Bank Staff Working Paper, No. 408 (July, 1980), passim.

between production per capita and human welfare may not be as useful as had been thought.

Resource Endowment Complex

Eight energy, land, and area indicators constitute this complex. There is an extremely high similarity (correlations greater than .800) among the aggregate energy potential indicators and of these with total area (and GNP to a lesser degree). However, there is a sharp distinction visible in the correlations between the aggregate measures and the per capita measures. Moreover, arable land per capita and percent of land arable do not correlate highly with the other indicators, although the latter does relate significantly to arable land per capita agricultural population (.549 for 146 countries). The highest correlation (.624 for 88 countries) for any of these resource endowment indicators with GNP per capita involves non-renewable energy potentials per capita, an artifact, presumably, of the "OPEC effect." Although it appears that resource endowments have been of relatively modest importance in the history of modern economic development, this may not be so in the future.[7]

Accesibility and Connectivity Complex

Nine indicators of road, rail, and freight, and vehicle densities per capita and per unit area and length are largely uncorrelated. Transportation network needs seem simply non-generalizable among the world's array of large, small, flat, mountainous, and densely and sparsely populated nations. Except for motor vehicles per capita, which is treated as a technological indicator above, moreover, there is surprisingly little correlation with GNP per capita. (Road length per capita has the highest correlation, at .316 for 178 countries.) The one exception to the lack of intercorrelation in this complex is the correlation of .804 (120 cases) between rail freight ton-kilometers per capita and rail freight per unit of rail length, which is in part a function of the paucity of rail systems in a large number of countries and reflects sparse road networks in most Socialist countries.

External Relations Complex

The nine indicators in this set composed of trade turnovers, composition of exports, external debt, and international mail flows are similarly intercorrelated. Trade turnover per capita has its highest association (.656 for 169 countries) with the distribution

---

7. N.S. Ginsburg, "Natural Resources and Economic Development," *Annals,* Association of American Geographers 47, no. 3 (September 1957), pp. 197ff.

of exports to industrialized countries as a proposition of total exports. Export dependency on raw materials has a lower correlation (.527 for 123 countries) with the index of export concentration than might have been expected from UNCTAD pronouncements; and although trade turnover per capita has a moderately significant correlation with GNP per capita (.607 for 169 countries), growth in trade in the past decade is less strongly associated (correlation of .339 for 152 countries) with GNP per capita levels.

Policy Issues

The indicators employed in this study also may be grouped around key global and national policy issues. Eight such issues are discussed briefly below as a further guide to the interpretation of patterns of development which are dealt with elsewhere in the study.

Food and Population

Ten indicators relate directly to the problem of population growth and distribution and to national agricultural capacities. We have already observed that population growth among the world's countries has a significant positive correlation with agricultural population growth (.675 for 147 countries), whereas agricultural population growth is negatively correlated with percent of urban population (-.650 for 121 countries). The proportion of total population that is agricultural is naturally highly associated with the contribution of agriculture to GDP (.805 for 129 countries); it has negative associations with both wheat yields (-.559 for 95 countries) and paddy yields (-.483 for 108 countries). Surprisingly, perhaps, the growth of overall agricultural production, over two periods, has no significant association with any indicator of grain yield. Less surprisingly it is not associated with GNP per capita and the growth of GNP per capita. Clearly, the relationships between food-production potentials and population distribution are complex. On the other hand, it also is clear that the larger the agricultural sector, either in production or population terms, the poorer the country.

Energy and the "OPEC Effect"

Energy consumption and growth in aggregate and per capita terms are viewed by many observers as a significant indicator of national development; and high energy consumption and rapid growth is regarded as a sine qua non for economic growth and modernization. The "price revolution" in oil, however, whereby the international price quadrupled from 1972 to 1978 due to the oligopolistic behavior

of the oil-exporting countries (OPEC), has had profound effects on
most countries, their balance of payments, and their trade and
growth prospects. The economies of the industrialized countries
have experienced depression, especially in the energy-intensive
leading sectors.[8] Inflation, in part attributable to the "OPEC
effect," has in turn reduced their capacity to trade with and extend
aid to the developing countries. The non-oil producing countries
have experienced, on the whole, even worse repercussions, as the
cost of oil imports has risen and forced other imports, some of them
equally crucial to development, to be reduced accordingly. The OPEC
countries, for the most part, have experienced vast increases in
their GNP per capita and the financial resources available to their
governments for development and defense expenditures, but this is
only slowly and variably translated into palpable economic and
social advances.

Thus, the growth of GNP per capita from 1970 to 1975 has shown
virtually no correlation (.170 for 149 countries) with the growth of
commercial energy consumption per capita; and trade growth from 1965
to 1975 has a lower correlation with the growth of commercial energy
consumption per capita (.454 for 144 countries) and GNP per capita
in the 1970s (.386 for 151 countries) than otherwise would be
expected. Whereas recent gross energy consumption has a significant
correlation with total trade turnover (.692 for 143 countries), it
is less highly correlated with total population (.471 for 143
countries). Inflation, as indicated by the consumer price index
change from 1970 to 1975, has no significant correlation with any
development indicator considered in this study. To the extent that
inflation is associated with an energy crunch, it appears to have
affected most countries, if not equally, at least irrespective of
their energy consumption potential and trade and economic levels.

Strength

The comparative power of a country in international dealings
may be measured in terms of such indices as gross product,
agricultural land, trade, and energy potentials, and/or in terms of
the quality of the population (e.g., literacy, calories per capita)
and the effectiveness of its transportation system. However, high
intercorrelations among indicators on these subjects are less common
than might be expected. Gross national product is highly correlated
with total trade turnover (.832 for 169 countries) and with total
energy potential (.731 for 125 countries), but not significantly

---

8. See W.W. Rostow, *The World Economy: History and Prospect* (Austin:
University of Texas Press, 1978), pp. 303-4, passim.

with development of the transport system or quality of the
population. Calories per capita and literacy are highly correlated
(.725 for 146 countries), but neither is associated with product or
with land and energy resource potentials. They are more closely
associated with road and rail length densities, with correlations
ranging from .336 to .625.

Cohesion

The ability of a society to move as a unit along a chosen path
of development should be associated, it might be argued, with a high
degree of interaction, with the density of its transportation
network, its degree of urbanization, its population and
communication density, and the literacy of the population. GNP per
capita in fact is highly associated with vehicles per capita (.789
for 169 countries) and daily newspaper consumption per capita (.725
for 128 countries). It is less highly correlated with radios per
capita (.487 for 170 countries). However, these various measures of
potential for cohesion are less highly intercorrelated than one
might expect. The proportion of population urban correlates highly
only with literacy (.632 for 134 countries). Urban growth has no
significant correlation with the other measures. Urban primacy,
which might have been expected to have significant negative
correlation with measures of accessibility, literacy, and
consumption of communications media, does not. Physical
accessibility is not significantly associated with propinquity as
measured by urbanization and population density (with the exception
of rail-length density which shows a correlation coefficient of .841
for 186 countries with population density) and with the exchange of
ideas through radio and the press.

Growth

Although the areal distributions among countries of several
economic growth indicators are more similar to one another than to
the distributions of other development indicators (see below,
chapter II), they are nevertheless not highly associated among
themselves or with other growth indicators included in this study.
Surprisingly, population growth, viewed by many as the bane of
economic development, seems to be largely neutral. It has no
significant correlation, and no strongly negative one, with measures
of agricultural production growth, GNP per capita growth, trade,
investment, primary school enrollment, and commercial energy
consumption per capita growth.

Agricultural population growth, highly associated with total population growth among the world's countries, as noted, is not negatively associated with growth of GNP per capita (agricultural population growth 1965 to 1975, with GNP per capita growth 1960 to 1975: .439 for 141 countries, and with GNP per capita growth 1970-75: .439 for 141 countries). Gross domestic investment growth from 1965 to 1973 has less consequential association with GNP per capita growth and trade growth that some might expect (GNP per capita 1960-75: .464 for 125 countries; GNP per capita 1970-75: .451 for 125 countries; trade 1965-75: .388 for 119 countries); and there is virtually no correlation between investment growth and the growth of population and agricultural production in these periods. The question of how the gap between the rich and poor countries might be narrowed, in short, is provided with few answers from the data here assembled.

Another approach to the growth issue, of course, is to consider conditions or factors which permit, encourage, or bring it about. Papanek proposes five factors believed to "play an important role in the rate of investment and in the efficiency with which it is used and, therefore, in determining the rate of economic growth."[9] They are: (1) the natural resource endowment, (2) the stock of capital in place, particularly in industry and infrastructure, (3) the human resource endowment, (4) the flow of aid, and (5) the development strategy of the national government combined with its effectiveness in implementing it. This sensible approach closely conforms to the order of functional complexes of development indicators as described above. Although it is obvious that great variability attaches to levels of attainment across such wide categories of national development condition, there is some argument for a synthetic measure of development which may portend growth and which combines measures across the quantifiable factors he suggests. This possibility has been pursued, and a measure of growth potential (GROPOT) has been computed and its distribution discussed in chapter III.

Dependency/Autarky

Countries may be more or less dependent on one another or the international system in the degree to which, inter alia, they trade, depend on vulnerable exports, have high population-to-resources ratios, and poorly developed advanced educational and industrial

---

9. Gustav F. Papanek, "Development Theory at Three-Quarters Century," in *Essays on Economic Development and Cultural Change in Honor of Bert F. Hoselitz,* ed. Manning Nash (Chicago: University of Chicago Press, 1977), p. 281.

systems. Two measures of export vulnerability, percent of raw
materials in total exports and export concentration, are associated
with one another, but not significantly with exports as a percent of
GNP. They are negatively, though not highly, associated with total
trade turnover (-.431 for 130 countries and -.338 for 137 countries,
respectively). They have higher negative correlations with
manufacturing as a percent of GDP (-.600 for 116 countries and -.581
for 120 countries). Energy potentials, arable land per capita, and
agricultural production growth have no significant associations with
indicators of trade dependency or higher education. Higher
educational attainment is associated with economic modernization (at
above the .610 level) in the form of manufacturing in GDP, exports
as percent of GNP, and, in the case of third-level school
enrollment, with total trade turnover (.625 for 132 countries).

Need

Many of the indicators measure aspects of poverty, structural
vulnerability, or decay to which the richer countries are asked and
expected to respond with assistance. Although inflation, external
debt, total energy potential, and negative economic growth are not
highly associated with measures of economic and social welfare and
trade dependency, measures of these latter conditions are highly
enough intercorrelated to validate attempts to order developing
countries along some kind of composite measure of need for
assistance. GNP per capita, calories per capita, primary and
secondary school enrollment ratios, life expectancy, infant
mortality, and youthfulness of population have an average
intercorrelation, disregarding sign, of .666. Inclusion of the
indicators of export dependence on raw materials and export
concentration reduces the average intercorrelation just to .573.
Further exploration of scales of development, modernization, and
rural poverty, and therefore need, have been undertaken (see below)
as a means for helping determine need.

Action

Societies and their governments have recourse to a variety of
measures to promote development, many of which are unquantifiable
and/or controversial (such as leadership, international bloc action,
or cultural revolution). Others may be a matter of investment in
human and economic resources, which carry with them implicit
structural change. Most of the indicators considered in this study
reflect such investment at one time or another, which in most
countries of the world is directed and largely controlled by central

governments. However, actual gross domestic investment as a static percentage of GDP in 1973 and its growth from 1965 to 1973 are not highly correlated with measures of technological and social consumption, trade growth, or total international assistance receipts. On the other hand, there are significant associations among countries' abilities to provide social services, certain industrial inputs, and, where cases warrant them, extensive transport systems. They do this, moreover, with a highly variable reliance on foreign borrowing and acceptance of per capita economic assistance. Why this is so remains unclear, but it suggests that the courses of action available to societies and their governments may be even more constrained than had been thought.

CHAPTER II
A MULTIVARIATE ANALYSIS

A basic premise of this enquiry is that the developmental process is too complicated and "development" itself too subject to variation in definition to permit a simple ranking of countries along some single scale of problem and prospect, without obscuring the policy issues discussed above. An informed perspective on the developmental status of the world's nation-states requires consideration of a wide range of variables. On the other hand, some reduction of the infinite complexity of the idiosyncratic clearly is desirable. To this end, the multivariate statistical methods employed condense much of the variation among the set of indicators to a few basic scales. The global patterns displayed on these scales have been mapped so that differences among countries can be readily observed. Some additional composite indices of development also have been developed. These provide summed scores for groups of indicators of special interest (chapter III).

The primary multivariate analysis method employed in the study is principal components analysis. The characteristics of this method and the reasons for choosing it are discussed at length in an addendum by Grant Blank, but a few words of introduction here may be useful.[1]

The Factor Analysis

Factor analysis usually begins with a data set consisting of a number of related variables. The initial goal is to construct a small number of new variables from the original variables in such a way so that the constructed variables account for a large proportion of the variance in the original data. The constructed variables are called factors. There are usually many fewer factors than there were original variables; reductions of ten to one and more are not unusual. Since the number of variables in the data is referred to

---

1. A cluster analysis was also run. It proved useful for interpreting many of the bivariate relationships among variables, as reflected in the previous discussion on Complexes of Variables, but the analysis is not reported on at length since its multivariate explanatory powers duplicate the findings of the factor analysis.

as the "dimensionality" of the data, the prime characteristic of
factor analysis is often referred to as *reduction of
dimensionality*.[2] Discussions of factor analysis are often confusing
because there are two quite different statistical procedures called
factor analysis. The more precise names of the procedures are
common factor analysis and principal components analysis. (See the
discussion of Multivariate Methodology in Cross-National Research,
following chapter V, for a detailed comparison of these two
methods.) For exploratory work such as this, principal components
is a more appropriate choice. Throughout this monograph, references
to factors, factor analysis, factor loadings, and factor score
coefficients always refer to the output from principal components
analysis.

The factors generated by a factor analysis are statistically
unrelated (i.e., uncorrelated) to each other, but they are
correlated with the variables in the data set. These correlations,
called factor loadings, are crucial to interpreting the results of a
factor analysis. The higher the factor loading, the more closely
the variable is related to the factor. Interpretation of the
meaning of a factor usually consists of examining the variables
which have high loadings on that factor but small loadings on all
other factors. A good interpretation identifies a common theme that
is consistent across all those variables.

The data presented two major problems for the factor analysis.
First, countries frequently do not collect data on every variable of
interest. The resulting problem of missing data plagues every
researcher working with cross-national data. In this research, only
four variables and one country had no missing data. To compound the
problem, data were not missing in random patterns. Small nations
and poor nations tended to have a disproportionate amount of missing
data. Under these circumstances, statistical procedures will
inevitably be biased. This is a major reason why many analysts have
been forced to reduce their set of countries to a highly
unrepresentative number (such as 58 in the case of the (UNRISD).[3]

There are various options for handling missing data in
multivariate analysis. The choices here are not easy because there
is no way to completely avoid all of the problems caused by missing

---

2. R. Gnandesikan, *Methods for Statistical Data Analysis of Multivariate
Observations* (New York: John Wiley & Sons, 1975).

3. United Nations Research Institute for Social Development (UNRISD),
Contents and Measurement for Socio-Economic Development: An Empirical Inquiry,
(Geneva, 1970). See the essay on Multivariate Methodology in Cross-National
Research, following chapter V, for a summary of other studies and an extended
discussion of this issue.

data. At best, a researcher can minimize the problems that seem to be most severe and interpret the resulting statistics with caution, knowing that this is not the best of all possible worlds. In this study an attempt was made to balance the twin goals of obtaining data for as many nations as possible without losing too much of the statistical validity in the multivariate procedures.

This led us to a two-fold approach. For a number of variables and countries, the proportions of missing data were simply unacceptably high. In addition, the data set contained some variables which were close surrogates for other variables. On these bases, forty-two countries and fifty-eight variables were eliminated. For the remaining 143 countries and 67 variables, the most statistically sound method was to estimate values for the missing data. The actual technique used was a stepwise multiple regression of each missing value on the available predictors.[4]

The second major problem was that factor analysis is strongly influenced by the scale of measurement. The units in which cross-national data are measured varied widely. On one hand, many variables measuring percent change had a theoretical range between zero and one hundred. On the other hand, there were variables like 1975 GNP with a range of $40 million (for Tonga) to $1.52 trillion (for the United States)--a range of 1 to 38,000. Furthermore, almost every variable was highly skewed. Skewness coefficients ranged from -2.0 (for agricultural production growth, 1970-74) to 11.0 (for consumer price index change).

Since formal hypothesis-testing was not an objective, the data did not have to be normally distributed or have constant variance. However, empirical transformations were developed to make each variable as symmetric as possible. The logarithmic transform was the most common. The reduction in skewness made the means and other statistics--such as variances and covariances--used by the multivariate methods much more accurate as measures of central tendency and spread. The variables further standardized by conducting all multivariate analyses from the correlation matrix. For the factor analysis, this had the effect of giving all variables equal weight.

---

4. This method is much superior to the use of pairwise deletion when constructing correlation or covariance matrices. See James W. Frane, "Some Simple Procedures for Handling Missing Data in Multivariate Analysis," *Psychometrika* 41 (1976): 409-415 and E. M. L. Beale and R. J. A. Little "Missing Values in Multivariate Analysis," *Journal of the Royal Statistical Society,* series B, 37 (1975): 129-145 for a general discussion of these issues and the advantages of this particular technique.

TABLE 2

PRINCIPLE COMPONENTS ANALYSIS: VARIABLES

	Factor 1	Factor 2	Factor 3	Factor 4	Factor 5
1. ALAB % of Economically Active Population in Agriculture	-0.933	0.0	0.0	0.0	0.0
2. P-APOP75 % of Population in Agriculture	-0.928	0.0	0.0	0.0	0.0
3. GNP75-C Gross National Product Per Capita	0.913	0.0	0.0	0.0	0.0
4. VHCL-C Motor Vehicles Per Capita	0.900	0.0	0.0	0.0	0.0
5. CEN75-C Commercial Energy Consumption Per Capita	0.897	0.0	0.0	0.0	0.0
6. DOC-C Physicians and Dentists Per Capita	0.879	0.0	0.0	0.0	0.0
7. EGEN-C Electricity Generation Per Capita	0.878	0.0	0.0	0.0	0.0
8. CMOT-C Commercial Motor Vehicles Per Capita	0.872	0.0	0.0	0.0	0.0
9. LIFE Life Expectancy at Birth	0.868	0.0	0.0	0.0	0.0
10. P-ACDP Agriculture % of GDP	-0.866	0.0	0.0	0.0	0.0

#	Variable					
11.	SEC70 — Secondary School Gross Enrollment Ratio	0.850	0.0	0.0	0.0	0.0
12.	P-CEN — Commercial Energy Consumption % of Gross Energy Consumption	0.847	0.0	0.0	0.0	0.0
13.	RADO — Radio Ownership Per Capita	0.841	0.0	0.0	0.0	0.0
14.	FCON-CA — Fertilizer Consumption Per Capita Agricultural Population	0.840	0.0	0.0	0.0	0.0
15.	DNEW-C — Daily Newspaper Circulation Per Capita	0.836	0.0	0.0	0.0	0.0
16.	LIT — Adult Literacy Rate	0.837	0.0	0.0	0.0	0.0
17.	TUR75-C — Trade Turnover Per Capita	0.822	0.0	0.0	0.0	0.0
18.	P-URB70 — Urban % of Total Population	0.820	0.0	0.0	0.0	0.0
19.	STEL-C — Steel Consumption Per Capita	0.817	0.0	0.0	0.0	0.351
20.	TRAC-CA — Tractors Per Capita Agricultural Population	0.811	0.0	0.0	-0.299	0.0
21.	TER — Third Level Gross Enrollment Ratio	0.793	0.394	0.0	0.0	0.0
22.	GREN-C — Gross Energy Consumption Per Capita	0.769	0.0	0.0	0.0	0.0

TABLE 2, *Continued*

	Factor 1	Factor 2	Factor 3	Factor 4	Factor 5
23. CAL-C Total Calories Per Capita Per Day	0.755	0.0	0.0	−0.402	0.0
24. TRAC-ARB Tractors Per Km^2 Arable Land	0.753	0.0	0.0	−0.458	0.0
25. VHCL-ROA Motor Vehicles per Km Road	0.749	0.0	0.312	0.0	0.0
26. FCON Fertilizer Consumption per Km^2 Arable Land	0.734	0.0	0.0	0.0	0.0
27. MAIL-C International Mail Sent Per Capita	0.724	0.0	0.0	0.0	0.0
28. PRI70 Gross Primary Enrollment Ratio	0.722	0.0	0.0	0.0	0.0
29. APOP-G Agricultural Population Growth 1965-1975	−0.667	0.0	0.0	0.446	0.0
30. P-UPOP % Population in Cities over 100,000	0.601	0.501	0.0	0.0	0.0
31. P-MGDP Manufacturing % of GDP	0.524	0.323	0.0	−0.353	0.0
32. POP75 Total Population	0.0	0.941	0.0	0.0	0.0

#	Variable					
33.	GNP75 Total Gross National Product	0.444	0.854	0.0	0.0	0.0
34.	CEN75 Total Commercial Energy Consumption	0.518	0.804	0.0	0.0	0.0
35.	TUR75 Total Trade Turnover	0.558	0.714	0.0	0.0	0.0
36.	RAILFDEN Rail Freight ton/km	0.0	0.697	0.0	0.0	0.0
37.	AREA Total Area	0.0	0.692	-0578	0.0	0.0
38.	TRNS-C Air plus Rail Passenger-Km Per Capita	0.348	0.637	0.0	-0.375	0.0
39.	RAILDDEN Rail Density Km/Km^2	0.317	0.560	0.266	-0.307	0.0
40.	POPDEN75 Population Density	0.0	0.0	0.904	0.0	0.0
41.	P-ARB1 % Arable Land of Total Area	0.0	0.0	0.826	-0.262	0.0
42.	BIOE-C Biomass Energy Potential Per Capita	0.0	0.0	-0.813	0.0	0.0
43.	ROADLDEN Road Density Km/Km^2	0.508	0.0	0.638	-0.398	0.0
44.	POP-G Population Growth 1965-75	-0.343	0.0	0.0	0.653	0.0
45.	YUTH Youthfulness of Population % 14 or Under	-0.562	-0.253	0.0	0.632	0.0

24

TABLE 2, *Continued*

	Factor 1	Factor 2	Factor 3	Factor 4	Factor 5
46. EXCON Export Concentration	0.0	-0.371	0.0	0.618	0.0
47. WEAT Wheat Yield	0.401	0.0	0.0	-0.594	0.0
48. GNPC60-G GNP Per Capita Growth 1960-75	0.274	0.0	0.0	0.0	0.797
49. GNPC70-G GNP Per Capita Growth 1970-75	0.0	0.0	0.0	0.0	0.788
50. CEN_CG Commercial Energy Consumption Per Capita Growth 1965-75	0.0	0.0	0.0	0.0	0.695
51. TUR-G Trade Turnover Growth 1965-75	0.0	0.0	0.0	0.0	0.642
52. P-GDl Gross Domestic Investment % of GDP	0.436	0.0	0.0	0.0	0.631
53. ARBL-C Arable Land Per Capita	0.0	0.0	0.0	0.0	0.0
54. ARBL-CA Arable Land Per Capita Agricultural Population	0.355	0.0	0.0	0.0	0.0
55. P-INDIM Trade with Industrial Countries % of Total Trade	0.0	0.0	0.0	0.0	0.0

56. P-EXP Exports % of GNP	0.0	-0.265	0.0	0.0	0.0
57. RAILL-C Rail Length Per Capita	0.0	0.559	0.0	0.0	0.0
58. PRIM Urban Primacy	0.0	0.436	0.0	0.0	0.0
59. URB-G Urban Growth 1960's	0.0	0.0	0.0	0.0	0.0
60. ROADL-C Road Length Per Capita	0.470	0.0	-0.465	-0.381	0.0
61. HGEN-C Hydro-Electricity Generation Per Capita	0.278	0.475	0.0	0.0	0.0
62. RAILF-C Rail Freight Per Capita	0.252	0.341	-0.281	0.0	0.0
63. PADY Padi (rice) yield	0.476	0.342	0.0	0.0	0.0
64. RAW % Raw Materials in Total Exports	-0.371	-0.355	-0.274	0.485	0.0
65. PRI-G Primary Enrollment Ratio Growth 1960-70	-0.450	0.0	-0.348	0.0	0.0
66. MORT Infant Mortality	-0.387	0.0	0.0	0.348	0.0
% of Variance Explained	36.23	12.17	6.51	6.46	4.95

NOTE: 0.0 = .250 or less.

An inspection of the array of factor loadings in table 2 is quite rewarding. They are sorted on each of five factors. Insignificant factor loadings (less than 0.25) have been converted to zero to facilitate comprehension. There appears to be a clear and consistent separation between a "national development" or "modernization" component (Factor 1) from those representing what has been termed "national magnitude," "population density," "population growth," and "economic growth" (Factors 2 through 5, respectively).

For general comparisons of countries, *Factor 1* is clearly the most interesting. It explains a high percentage of the total variance, compares well with the "Technological Scale" identified in the original *Atlas,* and contains many of the variables considered in other multivariate studies.[5] It suggests that GNP per capita, the conventional measure of both wealth and of development is not as poor a measure of attainment as is sometimes proclaimed (see figure 10). Moreover, it lends itself to further analysis which, as discussed below, more correctly specifies where various countries can be located.

The other components support the notion that national development is not associated simply with per capita wealth; that population growth, in the short run at least, although inversely associated with national development, cannot be considered a strictly negative attribute of development; and that economic growth per capita since the 1960s has, with few exceptions, had little impact on current levels of development.

Five Perspectives on Development

The five scales termed Factors 1 through 5 may be regarded as offering differing perspectives on, or structures of, development and the developmental process as derived from data chiefly for the mid-1970s. They are displayed on figures 1 and 3-5. To assess the relative position of individual countries, factor scores were generated for the five factors. Factor scores for countries are standardized with a mean of zero and a standard deviation of one. Thus, the positive and negative values represent distance from the mean in standard deviations. A six-way classification of scores has been adopted, composed of two sets of three categories each above

5. E.g., UNRISD, ibid.; Bruce Russet, *International Regions and the International System* (Chicago: Rand McNally, 1967); C.L. Taylor and M.C. Hudson, *World Handbook of Political and Social Indicators* (New Haven: Yale University Press, 1972); Irma Adelman and C.T. Morris, *Economic Growth and Social Equity in Developing Countries* (Groningen: Rotterdam University Press, 1969); and a general review of these and others in David M. Smith, *Human Geography: A Welfare Approach* (London: Edward Arnold, 1977), pp. 201-240.

and below the mean. As shown on table 3, each category separated by a solid line conveniently contains an approximately equal number of countries. A somewhat similar taxonomy is used in the maps of individual indicators which appear elsewhere in this volume, thereby facilitating comparisons.

Factor 1

Factor 1 provides a scale of development for 143 countries in the mid-1970s. Entitled "Economic Development per Capita," it explains 36.23 percent of the variance and has high correlations with indicators reflecting productivity, quality of life, industrial consumption, small agricultural populations, urbanization, and accessibility. It is particularly useful in distinguishing the very rich from the very poor countries (figure 1). It is less satisfactory for discriminating among countries which might be considered "moderately developed" or "developing." The highest category of attainment on the scale has values ranging from 1.12 to 2.05 standard deviations from the mean. It combines Anglo-America and Australasia with Western Europe, Israel, Venezuela (which reflected an extreme OPEC effect in the mid-1970s), and several very small entities of a special nature.

This association illustrates the pitfalls, as well as the utility, of the method used to derive these synthetic scales. The basic problem is that countries which have similar values on one or several indicators are usually quite different when seen from another perspective. This can easily be seen from another perspective. This can easily be seen by looking at countries grouped together on the map of Factor 1 (figure 1). They are similar with respect to the variables included in Factor 1 but are quite different in other respects. The highest category of Factor 1 includes traditionally powerful, rich, welfare-oriented industrialized states, such as those of Western Europe, as well as island, city-state, and oil-rich countries whose economic bases, societal structures, and stages of development are quite different.

This issue is, of course, the basis for one of the objections to the use of GNP per capita as a measure of development. Whether this objection is significant depends on the purposes for which the classification is being made. In this case, this goal was to do an exploratory study of the state of the world economy in the late 1970s. To be sure, no single variable can describe all of the relevant aspects of an economic system. That is the major reason why so many different maps are displayed, not only of the five

TABLE 3

PRINCIPAL COMPONENTS ANALYSIS: COUNTRIES

(143 Countries, 66 Variables)

OBS	Country Name	Economic Development Per Capita (Factor 1) Factor Score	Magnitude (Factor 2) CN	FS	Population Density (Factor 3) CN	FS	Population Growth (Factor 4) CN	FS	Economic Growth Per Capita (Factor 5) CN	FS
1	Kuwait	2.05	USSR	2.55	HONG	2.21	KUWA	2.89	BAHR	3.73
2	Iceland	1.69	PRCH	2.41	BURU	2.17	IRAQ	2.74	ROMA	2.38
3	Canada	1.68	USA	2.35	MALT	2.07	BAHR	2.39	GABO	2.31
4	Bahamas	1.65	INDI	2.19	MRUS	1.98	VENE	2.23	IRAN	2.25
5	USA	1.63	CANA	1.66	BANG	1.72	IRAN	1.99	BOTS	2.03
6	New Zealand	1.54	INDO	1.67	GUAD	1.63	LIBY	1.90	LIBY	1.87
7	Sweden	1.42	JAPA	1.58	MART	1.59	JORD	1.80	SKOR	1.75
8	Norway	1.40	BRAZ	1.57	SING	1.58	KAMP	1.69	SAUD	1.69
9	Australia	1.39	IRAN	1.50	HAIT	1.50	ECUA	1.51	NYEM	1.57
10	Israel	1.38	PAKI	1.37	BAHR	1.50	BOLI	1.50	SING	1.50
11	Netherlands	1.36	MEXI	1.33	RWAN	1.40	SYRI	1.42	SWAZ	1.29
12	United Kingdom	1.35	WGER	1.32	TRIN	1.38	ZAIR	1.40	NKOR	1.21
13	Venezuela	1.32	FRAN	1.22	REUN	1.25	PERU	1.32	PAPU	1.20
14	Switzerland	1.27	UK	1.18	LEBA	1.21	CUBA	1.31	ALBA	1.20
15	Belgium	1.27	ITAL	1.16	SIER	1.18	CHIL	1.29	ALGE	1.14
16	Martinique	1.26	EGYP	1.15	BELG	1.13	COLO	1.20	TUNI	1.12
17	Denmark	1.25	MGRA	1.12	HUNG	1.08	S.A.	1.18	LESO	1.05
18	Singapore	1.22	SKOR	1.11	ROCH	1.07	BRAZ	1.03	GAMB	1.05
19	Malta	1.20	S.A.	1.11	NGRA	1.06	GUYA	1.01	GREE	1.03
20	West Germany	1.19	GANG	1.10	INDI	1.04	ZAMB	1.00	GNBS	0.96
21	Finland	1.18	POLA	1.08	TOGO	1.04	MART	0.97	MRIA	0.88
22	Austria	1.16	SAUD	1.07	CYPR	1.03	REUN	0.97	DOMR	0.78

23	France	1.15	ZAIR	0.97	EL S	1.02	DOMR	0.94	YUGO	0.78
24	Bahrain	1.12	TURK	0.93	SKOR	1.02	ALGE	0.94	THAI	0.77
25	Uruguay	1.11	SPAI	0.92	JAMA	0.99	GNEA	0.85	POLA	0.77
26	Japan	1.0?	PHIL	0.87	ITAL	0.97	INDO	0.84	ROCH	0.75
27	Surinam	1.04	COLO	0.86	WGER	0.96	PHIL	0.83	SPAI	0.68
28	French Guiana	1.02	ROMA	0.81	DENM	0.92	BAHA	0.76	PRCH	0.66
29	Cyprus	1.02	ARGE	0.74	PORT	0.90	SURI	0.74	BULG	0.66
30	Libya	0.98	ETHI	0.74	SRI	0.85	BELI	0.73	JAPA	0.65
31	East Germany	0.98	YUGO	0.73	U.K.	0.84	NGRA	0.73	FIJI	0.65
32	Guadaloupe	0.94	EGER	0.73	PHIL	0.83	CONG	0.70	TURK	0.63
33	Argentina	0.94	THAI	0.72	NETH	0.82	MLSA	0.69	HONG	0.62
34	Trinidad	0.89	CZEC	0.69	EGER	0.82	MRUS	0.68	CONG	0.62
35	Reunion	0.89	SWED	0.66	KAMP	0.79	MORO	0.66	FINL	0.55
36	Italy	0.88	VENE	0.66	GAMB	0.72	PNMA	0.63	MART	0.54
37	Guyana	0.87	IRAQ	0.65	THAI	0.71	EGYP	0.60	IVOR	0.53
38	Chile	0.85	BURM	0.65	ISRA	0.66	THAI	0.60	ECUA	0.51
39	Lebanon	0.85	PERU	0.64	CZEC	0.66	HOND	0.53	USSR	0.51
40	Spain	0.84	SUDA	0.60	CUBA	0.64	HONG	0.53	BRAZ	0.47
41	Czechoslovakia	0.84	AUSL	0.59	MLWI	0.63	EL S	0.52	NGRA	0.47
42	Ireland	0.81	NKOR	0.59	POLA	0.62	BURM	0.51	TANZ	0.46
43	USSR	0.79	MORO	0.58	UGAN	0.61	MEXI	0.50	MLSA	0.50
44	Panama	0.73	ROCH	0.54	SYRI	0.54	ARGE	0.49	ANGO	0.45
45	Belize	0.73	ANGO	0.51	SPAI	0.51	ISRA	0.47	PORT	0.42
46	Hong Kong	0,67	ALGE	0.51	FRAN	0.49	CANA	0.44	TOGO	0.38
47	Hungary	0.60	CHIL	0.48	PAKI	0.47	GUAT	0.41	MALT	0.35
48	South Africa	0.58	NETH	0.45	GNEA	0.44	SAUD	0.39	COST	0.33
49	Greece	0.58	MOZA	0.44	IRAQ	0.43	JAMA	0.38	FR G	0.31
50	Jamaica	0.56	BULG	0.42	ROMA	0.40	TUNI	0.37	IRAQ	0.29
51	Portugal	0.50	CUBA	0.42	DOMR	0,39	PAKI	0.34	MLWI	0.29

52	Peru	0.49	TABZ	0.39	KUWA	0.34	GUAD	0.33	INDO	0.28
53	Mauritius	0.49	AFGH	0.35	GHAN	0.34	LIBE	0.33	AUSA	0.23
54	Bulgaria	0.46	HUNG	0.35	INDO	0.32	IVOR	0.33	KENY	0.21
55	Poland	0.45	GHAN	0.29	TURK	0.30	BANG	0.32	MALI	0.19
56	Cuba	0.42	KENY	0.26	YUGO	0.29	BURU	0.28	ISPA	0.19
57	Jordan	0.41	BELG	0.26	BULG	0.28	LEBA	0.27	SYRI	0.16
58	Republic of China	0.41	MLSA	0.24	PRCH	0.27	PRCH	0.26	REUN	0.15
59	Costa Rica	0.39	SWIT	0.24	JAPA	0.25	TRIN	0.26	UPR	0.14
60	Mexico	0.38	FINL	0.24	JORD	0.22	FR G	0.25	NICA	0.12
61	Yugoslavia	0.35	GREE	0.23	UPR	0.22	TURK	0.24	JORD	0.11
62	Iraq	0.29	ZAMB	0.23	NEPA	0.22	GHAN	0.23	BAHA	0.10
63	Fiji	0.29	MADA	0.20	LESO	0.19	SOMA	0.21	HONG	0.07
64	Colombia	0.28	PORT	0.17	IVOR	0.19	FIJI	0.21	SOMA	0.05
65	Mongolia	0.26	AUSA	0.16	MLSA	0.18	HONG	0.16	SURI	0.02
66	Syria	0.16	SYRI	0.16	TUNI	0.18	USSR	0.06	BOLI	-0.00
67	Brazil	0.14	UGAN	0.16	GUAT	0.17	PARA	0.04	CAME	-0.02
68	South Yemen	0.13	NORW	0.16	MORO	0.15	SING	0.03	ETHI	-0.04
69	Ecuador	0.09	CAME	0.07	SWAZ	0.15	NICA	0.01	SUDA	-0.04
70	Nicaragua	0.06	RHOD	0.05	BENI	0.14	SKOR	0.01	AUSL	-0.06
71	Iran	0.03	DENM	0.04	GREE	0.14	ROCH	0.01	PARA	-0.06
72	Dominican Republic	0.00	ECUA	0.04	NKOR	0.09	SUDA	-0.00	CZEC	-0.07
73	Malaysia	-0.01	GUAT	0.38	BURM	0.07	SYEM	-0.00	MORO	-0.08
74	Romania	-0.01	BOLI	0.03	SWIT	0.05	COST	-0.01	JAMA	-0.08
75	Tunisia	-0.03	LIBY	0.03	AFGH	0.04	SWAZ	-0.03	MRUS	-0.11
76	Egypt	-0.12	SENE	-0.02	AUSA	0.04	LESO	-0.04	ITAL	-0.12
77	Paraguay	-0.13	DOMR	-0.02	S.A.	0.02	URUG	-0.08	PHIL	-0.14
78	Albania	-0.14	IVOR	-0.03	ECUA	0.01	KENY	-0.09	GUAD	-0.15
79	Congo	-0.14	TUNI	-0.09	ALBA	-0.01	USA	-0.14	S.A.	-0.15
80	Zambia	-0.16	HONG	-0.11	FIJI	-0.04	GABO	-0.15	NORW	-0.17

81	El Salvador	-0.17	ISRA -0.14	SENE -0.08	SENE -0.16	BENI -0.19
82	Gabon	-0.19	SRI -0.18	IRAN -0.13	TOGO -0.16	PNMA -0.19
83	Turkey	-0.21	NZEA -0.20	IREL -0.16	INDI -0.19	BELI -0.20
84	South Korea	-0.21	NEPA -0.21	EGYP -0.20	AFGH -0.20	EGER -0.20
85	Bolivia	-0.22	KUWA -0.23	USA -0.20	MRTA -0.21	FRAN -0.21
86	Morocco	-0.26	GNEA -0.25	CAME -0.22	MLWI -0.23	NETH -0.22
87	Philippines	-0.28	LIBE -0.27	GMBS -0.22	CAME -0.24	CANA -0.25
88	Algeria	-0.28	MALI -0.32	MEXI -0.25	MADA -0.30	RWAN -0.29
89	Rhodesia	-0.31	MRTA -0.32	NIGR -0.26	ETHI -0.32	GUAT -0.29
90	Honduras	-0.36	HOND -0.36	URUG -0.27	MOZA -0.35	EL S -0.29
91	Guatemala	-0.39	IREL -0.36	NYEM -0.28	NKOR -0.37	IREL -0.30
92	Equatorial Guinea	-0.40	EL S -0.44	ETHI -0.30	RWAN -0,39	COLO -0.30
93	North Korea	-0.49	NIGR -0.44	COST -0.31	BENI -0.39	TRIN -0,32
94	Saudi Arabia	-0.49	CHAD -0.45	CHIL -0.32	AUSL -0.41	HOND -0.32
95	Sri Lanka	-0.50	UPR -0.46	MADA -0.37	SRI -0.42	CUBA -0.33
96	Liberia	-0.55	NYEM -0.48	EQ G -0.38	MALI -0.46	VENE -0.37
97	Ghana	-0.56	SING -0.49	CENT -0.43	EQ G -0.47	MEXI -0.37
98	Swaziland	-0.65	CONG -0.50	HOND -0.45	NIGR -0.48	LEBA -0.38
99	Ivory Coast	-0.67	GABO -0.50	ARGE -0.51	UGAN -0.48	SIER -0.39
100	Thailand	-0.71	SYEM -0.51	KENY -0.52	CYPR -0.49	CHAD -0.43
101	Kampuchea	-0.75	HAIT -0.52	BELI -0.55	RHDO -0.49	PERU -0.44
102	Senegal	-0.79	PAPU -0.52	PNMA -0.55	TANZ -0.49	WGER -0.44
103	Pakistan	-0.85	URUG -0.52	ALGE -0.59	LAOS -0.49	LIBE -0.45
104	Papua New Guinea	-0.86	MLWI -0.54	COLO -0.66	NYEM -0.52	GUYA -0.46
105	Somalia	-0.86	KAMP -0.54	RHOD -0169	GAMB -0153	DENM -0.49
106	Laos	-0.88	SIER -0.56	NICA -0.71	SPAI -0.57	ARGE -0.49
107	Kenya	-0.91	LAOS -0.57	CHAD -0.71	SIER -0.57	EGYP -0.49
108	Burma	-0.92	NICA -0.60	LIBE -0.72	ALBA -0.64	NEPA -0.51
109	Indonesia	-0.97	MONG -0.61	MALI -0.73	CHAD -0.67	SWIT -0.52

110	Cameroon	-0.99	BURU	-0.62	VENE	-0.78	UPR	-0.68	ZAIR	-0.53
111	Zaire	-0.99	LEBA	-0.66	ZAIR	-0.83	HAIT	-0.75	ICEL	-0.54
112	Gambia	-1.03	TOGO	-0.67	TANZ	-0.83	NZEA	-0.75	BELG	-0.55
113	Angola	-1.06	ALBA	-0.67	LIBY	-0.85	CENt	-0.82	MONG	-0.57
114	Central African Rep.	-1.07	BOTS	-0.68	MOZA	-0.86	PORT	-0.85	CYPR	-0.58
115	Guinea Bissau	-1.08	COST	-0.68	SWED	-0.86	ROMA	-0.91	INDI	-0.61
116	Madagascar	-1.08	PNMA	-0.68	ZAMB	-0.87	PAPU	-0.91	PAKI	-0.65
117	Lesotho	-1.08	PARA	-0.69	SAUD	-0.86	MALT	-0.96	AFGH	-0.65
118	Benin	-1.08	RWAN	-0.73	BRAZ	-0.91	NEPA	-0.99	MOZA	-0172
119	Mauritania	-1.08	BENI	-0.73	BAHA	-0.95	JAPA	-1.02	CENT	-0.72
120	China, PR	-1.10	SWAZ	-0.74	PERU	-0.98	ANGO	-1.05	SRI	-0.74
121	Haiti	-1.13	JORD	-0.74	BOLI	-1.04	GREE	-1.11	ZAMB	-0.74
122	Mozambique	-1.14	JAMA	-0.83	SUDA	-1.04	U.K.	-1.17	SWED	-0.75
123	Sudan	-1.14	SURI	-0.86	USSR	-1.05	POLA	-1.22	SENE	-0.88
124	Botswana	-1.17	GUYA	-1.01	FINL	-1.08	YUGO	-1.25	GNEA	-0.91
125	Togo	-1.22	CENT	-1.07	LAOS	-1.09	EGER	-1.25	BANG	-0.95
126	Nigeria	-1.26	SOMA	-1.07	SOMA	-1.14	FRAN	-1.26	USA	-0.96
127	Guinea	-1.28	TRIN	-1.16	NORW	-1.20	GNBS	-1.29	NZEA	-0.98
128	India	-1.30	BAHR	-1.39	AUSL	-1.20	ITAL	-1.30	NIGE	-1.04
129	Afghanistan	-1.31	ICEL	-1.44	NZEA	-1.20	NETH	-1.30	HAIT	-1.08
130	Sierra Leone	-1.36	LESO	-1.47	SYEM	-1.31	BELG	-1.31	U.K.	-1.14
131	North Yemen	-1.40	FIJI	-1.50	PARA	-1.33	BOTS	-1.38	BURM	-1.14
132	Uganda	-1.44	BAHA	-1.60	GUYA	-1.34	HUNG	-1.46	EQ G	-1.16
133	Tanzania	-1.45	REUN	-1.61	CONG	-1.47	BULG	-1.50	RHOD	-1.23
134	Malawi	-1.49	MRUS	-1.67	CAMA	-1.58	CZEC	-1.55	BURU	-1.28
135	Chad	-1.52	MART	-1.79	PAPU	-1.68	NORW	-1.55	GHAN	-1.32
136	Niger	-1.57	EQ G	-1.87	ANGO	-1.81	WGER	-1.59	MADA	-1.35
137	Bangladesh	-1.64	MALT	-1.99	BOTS	-1.97	SWED	-1.59	UGAN	-1.35
138	Mali	-1.69	GAMB	-2.03	SURI	-2.02	DENM	-1.65	URUG	-1.39

139	Nepal	-1.72	GUAD	-2.08	MRTA	-2.05	FIML	-1.76	CHIL	-1.64
140	Burundi	-1.74	CYPR	-2.12	GABO	-2.14	ICEL	-1.94	LAOS	-2.32
141	Rwanda	-1.79	GMDS	-2.19	MONG	-2.27	IREL	-1.99	KUWA	-2.70
142	Ethiopia	-1.87	BELO	-2.30	ICEL	-2.37	SWIT	-2.01	KAMP	-3.07
143	Upper Volta	-1.96	FR G	-2.56	FR G	-2.56	AUSA	-2.04	SYEM	-3.86

N=143

NOTE: Solid lines divide each factor into categories used in figures 1, 3-19. The dotted lines divide the countries in Factor 1 into three categories: those above one standard deviation above the arithmetic mean (29 countries); those within one standard deviation of the mean, positively and negatively (82 countries); and those more than one standard deviation below the mean (32 countries). Country names are spelled out for Factor 1, but are abbreviated for the remaining four factors, using the codes used for computer purposes.

34

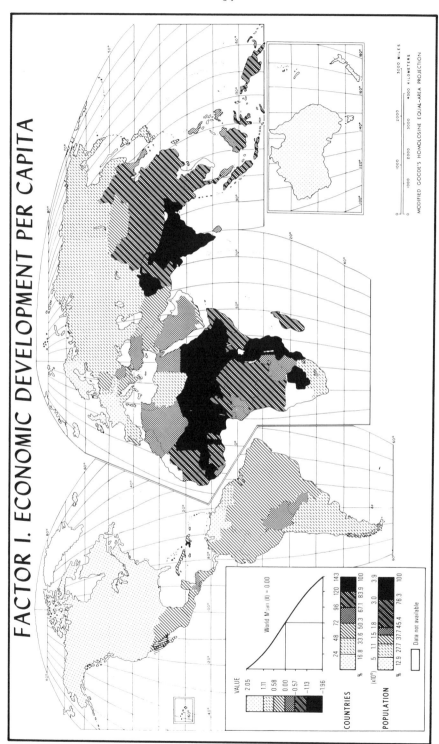

FACTOR I. ECONOMIC DEVELOPMENT PER CAPITA

Fig. 1

Factors, but also of many other variables related to economic growth.

The second highest category (0.58 to 1.11 standard deviations above the mean) neatly separates southern Europe, Japan,[6] the industrialized Eastern European Bloc, and wealthier low-latitude countries from both the wealthiest countries in the highest category and the band of countries which lie just above the mean.

At the lowest extreme of the global array, in contrast, the "poorest of the poor" stand out distinctly in sub-Saharan Africa and the Indian sub-continent; and the next higher category --the penultimately "poorest of the poor" nations--covers areas contiguous with those of the poorest countries, almost exclusively in Sub-Saharan Africa and parts of East and Southeast Asia.

When one concentrates on the two intermediate categories of countries which straddle the mean, it appears that most of Latin America lies above it, with the exception of two enclaves of relative poverty, one in Central America and one in parts of Andean South America. The Near East, except for certain OPEC countries which, with Lebanon and Israel, fall into the higher categories, also lies athwart the mean. Several Sub-Saharan African countries, with small populations and efficient export-oriented economies, appear in the category just below the mean, as do the two more prosperous of the Southeast Asian countries after Singapore: Malaysia and the Philippines.

The statistical distribution of values on Factor 1, as standardized scores, among the 143 countries (see table 3) suggests an alternative three-way classification of countries: (1) those above one standard deviation above the mean (29 countries); (2) those within one standard deviation from the mean, positively and negatively (82 countries); and (3) those more than one standard deviation from the mean (32 countries).[7] These categories have not been mapped, but their patterns should be quite apparent in figure 1.

The *first category* includes the industrialized countries of the West and Japan, along with some very small countries with peculiar development characteristics (Uruguay, Cyprus, and Malta, for example) and the more prosperous of the OPEC members.

---

6. To be sure, within a few years, Japan would move into the highest category.

7. In the Factor 1 column, the boundries of these categories are shown by a dotted line.

The *third category,* the unambiguously poor countries, consists of the populous low-income countries of Asia, including China, and the small poor countries of Africa, plus Haiti and North Yemen.

The large intermediary *second category* presents a range of countries where much progress has been made in recent years and in which there is the greatest heterogeneity of developmental characteristics. This group may be disaggregated in the following way.

The 22 countries within the range of values .50 to .99 above the mean are largely those whose development has tended to lag behind earlier Western models, as in the cases of southern Europe and certain island nations, and the Eastern European Bloc, but they still merit the accolade "developed." The 21 countries remaining above the mean and the 22 countries between 0 and -.50 below the mean represent combinations of natural resources, development policies, and international development-policy attention in recent decades such that they may be described broadly as "effectively developing." These 43 countries in the middle range (-.50 to +.50) include most of Latin America, the eastern and southern Mediterranean littoral, Malaysia and the Philippines in Southeast Asia, and the comparatively resource-rich countries of south-central Africa.

In contrast are those countries below the middle range, which fall between -.50 and -1.0 of the mean, about which there is increasing doubt as to their capabilities for sustained growth. It is among these 49 countries, and increasingly so as their distance from the world mean increases, that development policies and development assistance assume the guise of charity. This is not to say that an international policy of triage is either statistically or morally justified, for in most of these countries some real economic growth and improvement of welfare, in the absence of natural calamities, has taken place, but rather that the overwhelming effects of poverty forces concentration on human welfare in its most basic terms. Such concentration means that foreign assistance, for example, consists largely of subsidies in support of basic livelihood, welfare, and the most elemental forms of infra-structure, rather than investments which result directly in increases in human productivity.

The Disaggregation of Factor 1

One of the disadvantages of a principal components analysis is that a component with high "explanatory" power, like the first component, Factor 1, may obscure important variations within the population with which it deals. An examination of groupings of variables in a cluster analysis and the causal relationships between these groups and countries and regions, suggested that the first component was far too general and could be further refined.

As a result, a separate factor analysis was run on the 31 variables which were most highly associated with Factor 1. The sorted, rotated components that resulted from that analysis appear in table 4. As one can see, two, and only two, very strong components emerged from the analysis, accounting for over 77 percent of the variance in the variables. The most important data in that table appear in the right-most column labelled "Absolute Difference between A & B." This column is calculated as the absolute value of the difference between the factor loading on component A and the loading on Component B. The variables in the table have been ordered so as to highlight the difference between the two components. The procedure followed had two steps. First, the variables were divided into two groups: the upper group in the table consists of the variables which load most highly on Component A; the lower group loads higher on Component B. Second, within each group the variables were sorted on the difference between their loading on Component A and on Component B, as shown in the right-most column.

The components which emerge are not as "clean" as might have been expected, meaning that each component seems to include several variables with different or ambiguous meanings. Nonetheless, if one looks at the right-most column, which shows the variables that load unambiguously highly on one component, the pattern is clear.

Component A includes mostly variables measuring the delivery of government and private social services on a per capita basis. Hence, it includes variables such as literacy, two measures of school enrollment, doctors per capita, and urban population as a percent of total. The variable most out of place here is the percentage of manufacturing in gross domestic product, but it is not entirely so since the industrialized countries have a much more highly developed service sector. Less strongly related to this component are calories per capita, life expectancy, and a measure of agricultural productivity (fertilizer use). We will refer to this component as the *Social Services* component.

The second component B, includes primarily measures of transportation and communication per capita. Thus, it includes trade turnover per capita, vehicles per capita, vehicles per road mile, and commercial motor vehicles per capita. The other variable

TABLE 4

PRINCIPAL COMPONENTS ANALYSIS OF FACTOR 1

	Loading on Component		Absolute Difference between A & B
	A Social Ser.	B Transport	
Manufacturing % of GDP	0.84	0.19	0.64
School Enroll. 3rd Level	0.82	0.38	0.45
Literacy	0.83	0.42	0.41
Urban Pop Cities > 1 Mil	0.64	0.25	0.39
Physicians/100,000 Pop.	0.80	0.49	0.31
Fertil. Consump./Capita	0.72	0.42	0.31
School Enroll. Secondary	0.79	0.48	0.31
Life Expectancy	0.79	0.50	0.30
Calories per Capita	0.75	0.46	0.30
Fertil. Consump./Sq Km	0.78	0.51	0.27
Newspaper Circ./Capita	0.75	0.51	0.24
Agriculture Pop. Growth	-0.63	-0.43	0.20
Tractors/Capita Ag. Pop	0.70	0.53	0.17
Tractors/Sq Km Arable	0.65	0.50	0.15
Urban Pop. Census Def.	0.60	0.54	0.06
Electricity Gen/Capita	0.69	0.63	0.06
Radios per Capita	0.61	0.57	0.03
Commercl Enrgy % of Gross	0.63	0.60	0.03
Agric. Pop % of Total	-0.66	-0.66	0.00
Commercial Energy/Capita	0.67	0.68	0.01
Ag. Pop % of Ec Active	-0.66	-0.67	0.01
Gross Energy Consump/Cap	0.58	0.61	0.03
Steel Consum. per Capita	0.60	0.65	0.05
School Enroll., Primary	0.47	0.54	0.07
GNP per Capita	0.58	0.77	0.19
Vehicles per Capita	0.43	0.84	0.40
Agriculture % of Capita	-0.40	-0.80	0.40
Commercial Vehicles/Cap	0.39	0.83	0.44
Vehicles/Road Length	0.31	0.74	0.43
Mail Pieces per Capita	0.26	0.77	0.51
Trade Turnover per Cap	0.25	0.89	0.63
% of Variance	41.25	35.78	

NOTES: (1) Factor loadings after Varimax rotation and sorting.
        (2) Above the blank line in the table, variables load higher on
            the Social Services Factor than on Transportation

which loads highly on this component is the percentage of
agricultural products in gross domestic product. This makes sense
when you notice that it is negatively related and when you realize
that in most of the world agricultural production has a much simpler
division of labor than does industrial production and neither
requires nor makes use of much transport. This component may be
called the *Transportation* component.

Factor scores were calculated for each country on each of the
two components and plotted them in figure 2. The vertical axis
represents the Transportation component. The Social Services
component is plotted horizontally. Note that the two components are
essentially a measure of economic connectedness and a measure of
social development.

Two patterns are important here. First, notice the strong
regional patterns, particularly on the periphery of the figure.
This is consistent with the patterns that characterize, for example,
Gross National Product per capita and gross primary school
enrollment. It is also consistent with a multitude of other
research studies.[8] Second, notice how the regions arrange
themselves. The OECD nations are a very tight cluster, having a
small within-cluster variance. They form a well defined group that
ranks consistently high on both factors. This is similar to the
findings of Banks concerning the industrial countries in his
sample.[9] Among the less developed countries, the clusters are much
less tight. Although the regions are fairly distinct, the variation
within regions is large. This is also consistent with Banks.[10]

Low on the Transportation scale and on either side of the mean
of Social Services are found all the countries in South Asia, plus
neighboring Burma. Low scores on both scales are characteristic of
the "poorest of the poor," mainly sub-Saharan African countries.
Those with valuable exports rise higher on the Transportation
component within the cluster, even though many of them retain low
scores on the Social Services component.

8. Hollis B. Chenery and Peter Eckstein, "Development Alternatives for
Latin America," *Journal of Political Economy* 78 (1970): 966-1006; Hollis B.
Chenery and Moises Syrquin, *Patterns of Development,* (New York: Oxford University
Press); John K. Galbraith, "Underdevelopment: An Approach to Classification," in
Fiscal and Monetary Problems in Developing Countries, ed. David Krivine (New York:
Proceedings of the Third Rehovoth Conference), pp. 19-45; and International Bank
for Reconstruction and Development (World Bank), *World Development Report,*
(Washington D.C.: The World Bank, various years beginning in 1978).

9. Arthur Banks, "Industrialization and Development: A Longitudinal
Analysis," *Economic Development and Cultural Change* 22 (1974): 320-337.

10. Ibid.

THE DISAGGREGATION OF FACTOR 1

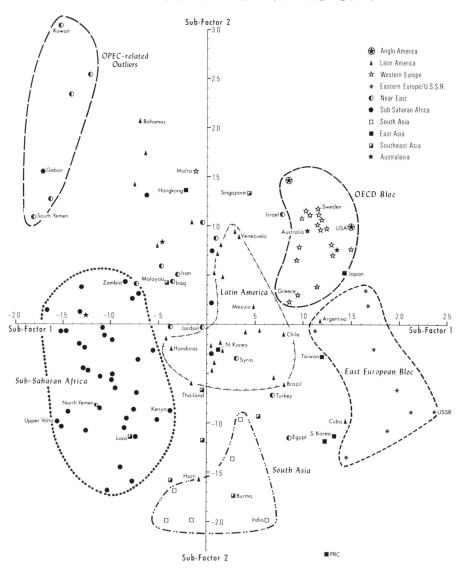

Fig. 2

Another peripheral cluster appears as an OPEC-related outlier in the upper left quadrant. The influx of petro-dollars and very large export trades with relatively small population have resulted in an extreme imbalance between Transportation scores and negative Social Services scores.

The point of this scatterplot is four-fold. First, it offers further confirmation of the need for disaggregation. This issue touches theoretical ground as well. The fact that regional similarities appear in the data on economic and social development suggests that these need to be accounted for in general discussion of economic development. Second, to follow the example of many studies and divide the world into two ("developed" and "less developed") or three categories does not make much theoretical sense. It is, of course, statistically possible to calculate a point estimate of the less developed countries, but what would it mean? The less developed countries are clearly *not* a homogeneous category. Analyses need to treat them in a much less aggregate form, categorizing by regions or some other relevant characteristic, than has heretofore been the case.

Third, the industrial OECD countries form the "best" of the regions in the sense that, on these two factors, they are the most homogeneous group of countries. In general, as levels of development decline, the regions become more spread out. The countries, though still an identifiable group, are less homogeneous. This suggests that when one disaggregates, one has to be very careful to recognize that groups consisting of less developed countries will have a larger variance than will groups of more developed countries and that point estimates of the characteristics of the former groups will be less meaningful. They will not summarize the state of the group as a whole very accurately. This suggests a need for caution in interpreting statistical results.

Finally, the obvious regional similarities do not obscure the fact that many countries fall outside the regional clusters that have been drawn on figure 2. See, for example, Taiwan and Argentina on the diagram. Regions are not a complete summary of the characteristics of any particular country. In the larger context of this paper, the point is that when a researcher is considering how to disaggregate data, regions should be an important candidate.

Factor 2

Factor 2, which explains 12.17 percent of the variance, associates variables measuring area, population size, GNP, trade

42

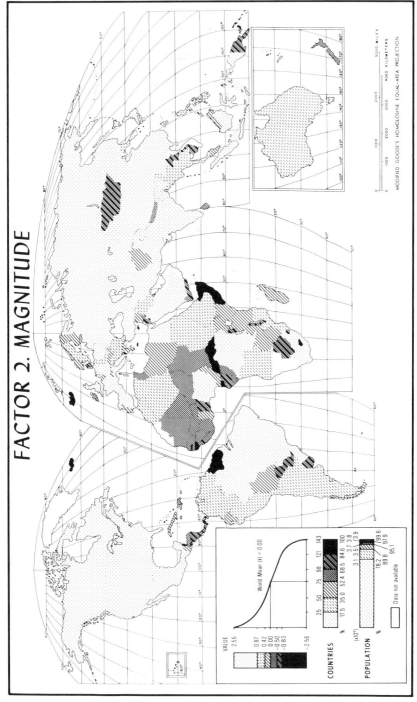

FACTOR 2. MAGNITUDE

Fig. 3

turnover, and energy consumption (figure 3). It appears to be a
measure of sheer size of country and economy, and therefore has been
given the name: "Magnitude." When interpreting the accompanying
map of factor scores, remember that the variables loading on this
factor do not have identical weights for each country. Thus, for
example, Saudi Arabia ranks in the first category chiefly on the
basis of a very high GNP and a large national area, whereas Eygpt is
in the same category primarily on the basis of a large population.
What figure 3 appears to do best is mark off smaller, poorer
countries from all the rest, although even here there are
exceptions, e.g., Iceland, small to be sure but certainly not poor.
Such countries account for only ten percent of the world's
population.

Factor 3

The patterns displayed by the mapping of Factor 3 on figure 4
appear to be more meaningful but still ambiguous. This factor,
which explains 6.51 percent of the variance, has been entitled
"Population Density," since that variable appears to dominate the
mix of variables. It highlights two types of countries both at the
lower end of the scale: those that are relatively wealthy and
populous but smaller on the one hand (as a number of European
countries), and on the other those that are very heavily populated,
whatever their size, and poorer. Population density itself is
associated on this measure with the proportion of land area arable
and with road density. Whereas national population density may, up
to a point, lend integrity to the socio-economic system, it
basically reflects (1) the ability of the land in poor countries to
support people through intensive agriculture and (2) the ability of
the economies of the wealthier small and medium-sized countries to
provide employment outside of agriculture that permits them to
purchase food from abroad if necessary. Western Europe and the
Asiatic Triangle region contain relatively densely populated
countries; the Americas and Australasia are, of course, distributed
across the entire range.

Factor 4

Population density and gross national product increase, of
course, with "Population Growth," the title given factor 4. Factor
4, which explains 6.46 percent of variance, associates medium-high
negative loadings of population growth on modernization (Factor 1)
and a positive loading on youthfulness of population and, perhaps
significantly, on export concentration. The patterns shown on

44

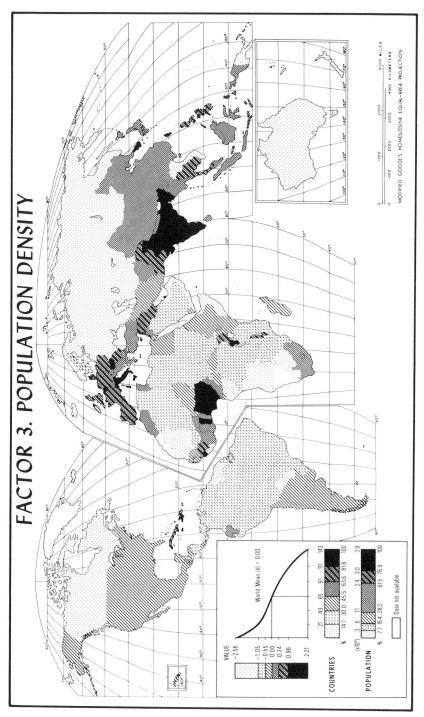

FACTOR 3. POPULATION DENSITY

Fig. 4

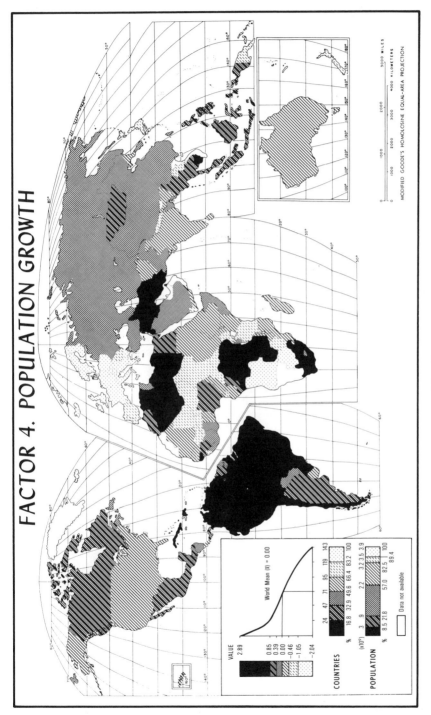

FACTOR 4. POPULATION GROWTH

Fig. 5

figure 5 in fact suggest that population dynamics are more likely to influence the distributions represented on figures 3 and 4 than on figure 1. The Western European countries as a group, not surprisingly, are characterized by extremely low factor scores, whereas in the Americas, especially in South America, countries have extremely high scores. With few exceptions, then, e.g., Canada and the U.S.S.R., the more highly developed states stand out clearly, but the developing countries are more varied. Note the low factor scores displayed by the Sahelian countries and some others in Africa, as well as the relatively low scores for India and Bangladesh.

Factor 5

Factor 4, which might be considered indicative of worsening human conditions in at least some of the poorer countries, may be contrasted with Factor 5, "Economic Growth per Capita," which explains 4.95 percent of the variance. The patterns on figure 6 reveal two modalities. Among the developed countries in the decade since 1965, the most developed grew the least rapidly except for the Eastern Bloc. The second rank of developed countries, however, have been catching up. These include Japan, South Korea, Taiwan, and the Comecon countries, all of which show high scores. Among the developing countries, high scores on this scale are achieved through combinations of export dynamism and economic efficiency which have more than kept pace with population growth. Thus, Brazil, Malaysia, the Koreas, and the less populous OPEC members rank very high; whereas the South Asian countries and those of Latin America (except Brazil and Ecuador) rank quite low. China's high score is something of an anomaly but reflects both a very low base level and the consequences of over two decades of peace and investment in infrastructure.

On the basis of these several distinctions, it appears that the traditionally developed countries are converging on a plateau of development associated in part with higher energy costs and the trade-off of employment for productivity, whereas the developing countries differ widely in their recent performance--varying from rapid growth either from a very low base line or via the OPEC effect, to what may be sustained growth up to some as yet unattained plateau (e.g., Brazil, China, Malaysia), to little or no progress at all. Comparing the patterns in figures 5 and 6 reveals the status of countries with respect to the tensions between population growth and growth of the economy. Most of South America is losing ground, as appears to be South Asia, whereas the Near East, excepting Eygpt, despite high population growth, is gaining.

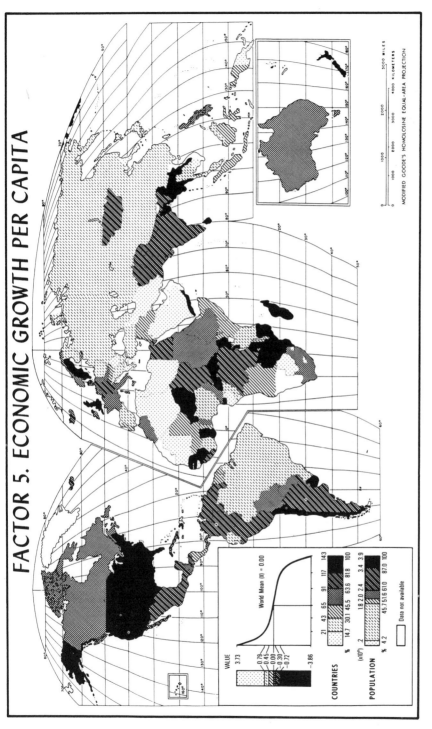

FACTOR 5. ECONOMIC GROWTH PER CAPITA

Fig. 6

CHAPTER III
ALTERNATIVES TO CONVENTIONAL MULTIVARIATE ANALYSIS

Countries as Variables

Another way to regionalize and group countries, using the data set constructed for multivariate analysis, is to transpose the matrix, treating the countries as variables and the indicators as observations. A factor analysis following this transformation yields orthogonal factors where the factor loadings display the degree to which each country is associated with each factor. The measures of development--what were the variables before--become the observations and are given equal weight. The thirty-three variables which loaded at or above .470 on Factor 1 in the original factor analysis and did not have higher loadings on other factors were chosen as the observations for such an analysis (refer to table 2).

The results, shown in table 5, provide yet another classification of countries on a composite measure of development. Two factors emerged after rotation which together account for an astonishing 99.7 percent of the total variance. Factor A is called *Poverty* and Factor B is *Wealth*. Each, it should be remembered, represents wealth or poverty not as defined in a conventional measure like GNP per capita but as measured largely, though not exclusively, by a combination of per capita indicators of product and consumption, with added indicators of economic and population structure. Of the 143 countries, 86 load more highly on Factor A and 57 more highly on Factor B.

The loadings on the two factors provide a picture of relative development for each country. In this table, one observes decreasing poverty down the column of Factor A as it is increasingly balanced by wealth on Factor B. Thus, the Sub-Saharan countries, beginning with Upper Volta, are joined in poverty by the countries of South Asia, then those of Southeast Asia, the Caribbean, and finally the Near East and Latin America on the ascending scale.

TABLE 5

ROTATED FACTOR LOADINGS, COUNTRIES AS VARIABLES

	Factor α Poverty	Factor β Wealth
Upper Volta	0.991	0.0
Chad	0.990	0.0
Central African Empire	0.989	0.0
Niger	0.986	0.0
Mali	0.985	0.0
Bangladesh	0.984	0.0
Mauritania	0.983	0.0
Rwanda	0.982	0.0
Nepal	0.980	0.0
Ethiopia	0.980	0.0
Madagascar	0.980	0.0
Tanzania	0.980	0.0
Burundi	0.978	0.0
Afghanistan	0.978	0.0
Malawi	0.976	0.0
Somalia	0.975	0.0
Togo	0.975	0.0
Gambia	0.974	0.0
Botswana	0.974	0.0
Cameroun	0.973	0.0
Gabon	0.973	0.0
Bissau	0.972	0.0
Laos	0.971	0.0
Lesotho	0.966	0.254
Senegal	0.965	0.252
Kampuchea	0.965	0.259
Papua New Guinea	0.964	0.264

TABLE 5, *Continued*

	Factor α Poverty	Factor β Wealth
Sudan	0.964	0.253
Guinea	0.963	0.0
Equatorial Guinea	0.962	0.267
Zaire	0.963	0.267
Liberia	0.962	0.268
Swaziland	0.962	0.270
Yemen, North	0.960	0.274
Angola	0.959	0.272
Uganda	0.959	0.272
Sierra Leone	0.957	0.0
Kenya	0.956	0.294
India	0.955	0.294
Zambia	0.954	0.301
Mozambique	0.953	0.301
Ivory Coast	0.945	0.287
Saudi Arabia	0.942	0.332
Haiti	0.937	0.348
Nigeria	0.936	0.339
Thailand	0.935	0.354
Indonesia	0.928	0.373
Ghana	0.920	0.390
Honduras	0.916	0.399
Rhodesia	0.908	0.417
Yemen, South	0.905	0.424
Guatemala	0.904	0.426
Turkey	0.898	0.438
Bolivia	0.895	0.445
Pakistan	0.894	0.447

TABLE 5, *Continued*

	Factor α Poverty	Factor β Wealth
Burma	0.894	0.446
Algeria	0.885	0.465
Dominican Republic	0.884	0.465
China, Peoples Republic	0.883	0.466
Morocco	0.880	0.474
Benin	0.877	0.478
Egypt	0.871	0.490
Albania	0.861	0.508
El Salvador	0.852	0.523
Nicaragua	0.851	0.524
Mongolia	0.848	0.529
Ecuador	0.844	0.536
Belize	0.837	0.539
Malaysia	0.834	0.552
Syria	0.830	0.557
Philippines	0.829	0.558
Sri Lanka	0.822	0.569
Korea, North	0.819	0.571
Iraq	0.816	0.576
Paraguay	0.816	0.576
Congo	0.814	0.560
Tunisia	0.809	0.578
Iran	0.808	0.588
French Guiana	0.796	0.605
Peru	0.792	0.609
Romania	0.779	0.627
Brazil	0.768	0.640
Korea, South	0.754	0.656

TABLE 5, *Continued*

	Factor α Poverty	Factor β Wealth
Yugoslavia	0.729	0.684
Mexico	0.729	0.684
Fiji	0.720	0.693
Costa Rica	0.706	0.708
Bahrain	0.698	0.712
Greece	0.691	0.723
Panama	0.678	0.734
Jordan	0.678	0.734
South Africa	0.675	0.734
Bulgaria	0.672	0.740
Colombia	0.666	0.746
Reunion	0.659	0.747
Cyprus	0.647	0.761
Mauritius	0.627	0.778
Poland	0.622	0.783
China, Republic of (Taiwan)	0.619	0.783
Portugal	0.559	0.828
Libya	0.540	0.838
Guyana	0.539	0.842
Cuba	0.525	0.849
Jamaica	0.525	0.851
Ireland	0.508	0.859
Bahamas	0.491	0.867
Chile	0.474	0.880
Venezuela	0.467	0.880
Surinam	0.458	0.888
Guadeloupe	0.453	0.889
Hungary	0.436	0.904

TABLE 5, *Continued*

	Factor α Poverty	Factor β Wealth
U.S.S.R.	0.427	0.904
Martinique	0.417	0.907
Spain	0.418	0.908
Iceland	0.373	0.923
Trinidad	0.375	0.926
Finland	0.360	0.932
Argentina	0.330	0.943
Lebanon	0.324	0.945
Uruguay	0.321	0.947
Italy	0.314	0.949
Japan	0.309	0.950
Czechoslovakia	0.292	0.956
Austria	0.258	0.965
Germany, East	0.0	0.968
France	0.0	0.970
New Zealand	0.0	0.971
Norway	0.0	0.971
Denmark	0.0	0.979
Kuwait	0.0	0.981
Israel	0.0	0.981
Australia	0.0	0.985
Sweden	0.0	0.986
Switzerland	0.0	0.987
Canada	0.0	0.987
Netherlands	0.0	0.988
Malta	0.0	0.988
Germany, West	0.0	0.989
Hong Kong	0.0	0.993

TABLE 5, *Continued*

	Factor α Poverty	Factor β Wealth
Belgium	0.0	0.994
Singapore	0.0	0.995
U.S.A.	0.0	0.996
United Kingdom	0.0	0.997
Percent of Variance Explained	57.7	42.0

Two blocs of countries emerge, which load at less than .800 on Factor B but have correspondingly high (greater than .600) loadings on Factor A. These 21 nations (read up from Republic of China on table 5) and several others close to them may be considered transitional from developing to developed status, without benefit of either the OPEC or extreme small-country effects. These transitional countries include Mexico, Brazil, Peru, Colombia, and Costa Rica in Latin America; the Republic of China (Taiwan) and South Korea in East Asia; and Yugoslavia, Romania, Poland, Bulgaria, and Greece in Europe.

The "World Standard" Development Indices

The factors described above and in chapter II have the virtues of being uncorrelated with one another and of possessing a remarkable clarity. However, the principal components approach (like most similar multivariate techniques) provides no control over which variables are included in any given factor, nor is there any way to control how each variable is weighted when factor scores are calculated.

To avoid these deficiences special scales of development were designed. The advantage of designing special composite measures is that by careful choice of variables they can be tailored to address specific issues. The chief value is strategic. As international development policy-makers have directed innovative strategic thinking toward alleviating the worst conditions of poverty and human degradation on the globe, greater refinement in the location and identification of such conditions than relative GNP per capita levels suggest is required.

Several reformulations of international development policy have been proposed in recent years.[1] These may be sectoral and specific, as in concentrating on increased food production and greater economic self-reliance; or they may concentrate on socio-economic targets and imply in many cases radical adjustments of national development policies—towards welfare and human resources development. A "Basic Needs" strategy would concentrate more investment, especially that originating in external assistance, on provision of essential consumption items (food, clothing, shelter), essential social services (potable water, sanitation, health care, and basic education), and productive employment. A

---

1. See, inter alia, OECD, *Review: Development Co-operation* (Paris, 1976), p. 22 ff.; Overseas Development Council, op. cit. and *World Bank, World Development Report 1980.* (Washington: IBRD, 1980); and Independent Commission on International Development Issues [The Brandt Commission], *North-South: A Program for Survival* (Cambridge: MIT Press, 1980).

policy called "Redistribution with Growth," as its title suggests, would redirect investment and resource transfers, with modification of wage policy, to benefit the rural poor, particularly the landless, and the destitute urban populations bound to the informal sector and a peasant ethos not yet replaced even after rural-urban migration.

What combinations of indicators will specify the distribution of need for such strategic policy choices among the developing countries? The discussion in the previous paragraph indicates that a heterogeneous variety of indicators is needed to measure the status and progress of almost any policy. GNP per capita used to identify the two dozen or so "least developed" countries, the "poorest of the poor," is too simple and not appropriate because it may be misleading. It is not sensitive enough to particular sectors of the economy or to particular issues (such as food or housing) to be useful.

To deal with this problem a set of "World Standard Distance Scales" has been created.[2] Distributions on two of these have been mapped and analyzed. The construction of these scales is simple. For each indicator, the world standard is established by averaging the values of the countries with the highest factor scores on *Factor 1*. After elimination of those countries displaying the OPEC effect and ones of less than 3 million population, 17 countries remained. The seventeen are Canada, USA, New Zealand, Sweden, Norway, Australia, Israel, Netherlands, United Kingdom, Switzerland, Belgium, Denmark, West Germany, Finland, Austria, France, and Japan. The value for each country in the data set is then converted to a percentage of this standard value, giving ranges of values from zero to somewhat above 100. An index is thus obtained which combines values on a number of indicators so derived by averaging them (but with cases [countries] removed which have more than half of the values missing).

The World Standard Distance approach provides a real measure of attainment among countries. The averaging process across a number of variables leading to a composite index assumes equal weight for all indicators. This assumption is, at least, readily comprehensible, although, to be sure, it is subject to some criticism; and it offers the advantage that index values can be

---

2. Eleven such indices were created, of which two were selected for mapping and discussion on the basis of their estimated analytical value. Those not mapped and analyzed here are: Quality of Population, Urban/Rural Population Distribution, Rural Poverty, Accessibility, Technology, Trade, Economic Structure, Growth, and Physical Quality of Life. It is hoped to present an analysis of these indices in a later paper.

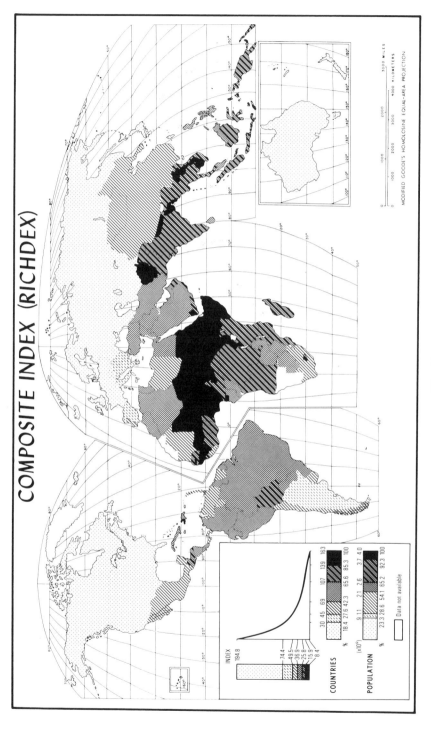

COMPOSITE INDEX (RICHDEX)

Fig. 7

calculated for more countries than is the case when complete sets of
data are necessary. The pattern of missing values for each country,
of course, means that the average scoring procedure gives different
emphasis to different indicators among many countries, but that
defect is partially mitigated by not calculating indices for
countries when more than half of the variables have missing values.
This problem is of minor significance when balanced against the
goals of attaining comparative measures applicable to as many
countries as possible and of recognizing the multi-dimensionality of
development among so many different nations.

The two special indices chosen for mapping and analysis here
are: the Composite Index (RICHDEX) and the Growth Potential Index
(GROPOT).

Composite Index (RICHDEX)

This measure, depicted on figure 7, is a composite of 42
indicators which theory suggests have direct relevance to
development (see table 6).[3] Its purpose is to define the distances
which separate 163 countries from the highest world standard of
development. Therefore, it may well offer a basis for the most
useful map of differential national develement across the world that
the data permit. The reader will not be surprised, however, that
the Composite Index patterns closely resemble those of Factor One as
displayed on figure 3.[4]

Thirty countries, accounting for 23 percent of the world's
population, are at levels of 74.4 percent or more of the World
Standard, including northwestern Europe, Eastern Europe, Japan,
Australia, Canada, the United States, and several island
micro-states. Countries reaching 49.5 - 74.4 percent of the World
Standard are clustered largely in southern Europe, but Israel,
Argentina, and the Republic of China (Taiwan) also are included.

Four categories of country scores below 49.5 percent have been
mapped, representing 72 percent of the 163 countries for which
scores are available and about 71 percent of human kind. These
developing countries are differentiated on this index with a kind of
refinement that may be especially useful in policy-making, as
contrasted with Factor One and with GNP per capita.

Thus, the countries in the 36.9 - 49.5 percent range,
including Mexico, Venezuela, Chile, Uruguay, Libya, South Africa,
Korea, and China, seem to require relatively little external

---

3. Iceland has an index score of 467, over twice that of the next highest
country. It was omitted from the legend on the grounds that its inclusion would
have made the legend indecipherable.

4. Note that the categories in figure 7 are highly subjective, and are not
based on any strict quantitative standard.

TABLE 6

COMPOSITE INDEX VARIABLES

1. Agriculture as Percent of Gross Domestic Product, 1973, Current Factor Cost

2. Agricultural Population, Percent of Economically Active Population

3. Agricultural Population,]975, Percent of Total Population

4. Calories per Capita per Day

5. Energy Consumption: Commercial Consumption as Percent of Gross Consumption, 1975

6. Energy Consumption per Capita: Commercial Consumption, 1975, Tons of Coal Equivalents per 100 Population

7. Energy Consumption per Capita: Electricity Generation, 1975, Kilowatt-Hours per Person

8. Energy Consumption per Capita: Gross Consumption, 1975, Tons of Coal Equivalents per 10 Persons

9. Energy Consumption per Capita: Hydroelectric Generation, 1975, Kilowatt-Hours per Person

10. Export Concentration, 1972, Hirschman Normalized Index

11. Exports of Raw Materials, 1973, Percent of Total Exports

12. Fertilizer Consumption, 1975, Kilograms per Square Kilometer of Arable Land

13. Fertilizer Consumption per Capita, 1975, Kilograms per 10 Agricultural Population

14. Infant Mortality, 1970, Deaths per 1,000 Live Births

15. Investment: Gross Domestic Investment as Percent of Gross Domestic Product, 1973

16. Literacy Rate, Adult, 1970, Percent

17. Newspaper Circulation per Capita, 1970, Daily Copies per Thousand People

18. Life Expectancy at Birth, 1970, Years

19. Manufacturing as Percent of Gross Domestic Product, 1973, Current Factor Cost

20. Paddy Yield, 1975, Tens of Metric Tons per Square Kilometer

21. Physicians and Dentists per 100,000 People, 1970's

22. Radios Ownership per Capita, 1973-74, Radios per Thousand People

TABLE 6, *Continued*

23. Rail Route Density, 1975, Kilometers per Thousands of Square
    Kilometers

24. Rail Route Length per Capita, 1975, Ton-Kilometers per
    100,000 People

25. Rail Frieght per Route Length, 1975, Thousands of Ton-
    Kilometers per Kilometer of Rail Route Length

26. Rail Freight per Capita, 1975, Ton-Kilometers per Million
    Persons

27. Road Network Density, 1975, Kilometers per Thousand Square
    Kilometers

28. Road Network Length per Capita, 1975, Kilometers per
    100,000 People

29. School Enrollment, Primary, Gross Enrollment Ratio,
    1970, Percent

30. School Enrollment, Secondary, Gross Enrollment Ratio,
    1970, Percent

31. School Enrollment, Third Level, Gross Enrollment Ratio,
    1970, Percent

32. Steel Consumption per Capita, 1975, Tons per Thousand
    People

33. Tractors per Capita, 1975, Tractors per Million
    Persons Agricultural Population

34. Tractors per Square Kilometer of Arable Land, 1975

35. Trade Turnover per Capita, 1975, US Dollars per Person

36. Urban Population, 1975, Percent in Cities over 100,000 People

37. Urban Population, 1970's, Percent of Total Population,
    Census Definition

38. Vehicles per Capita, 1974, Commercial Motor Vehicles per
    100,000 Persons

39. Vehicles per Capita, 1974, Motor Vehicles per Thousand People

40. Vehicles per Road Length, 1974, Motor Vehicles per Hundred
    Kilometers

41. Wheat Yield, 1975, Metric Tons per Square Kilometer

42. Youthfulness, 1970s, Percent of Population Age 14 or Less

assistance of a concessional or official sort: they are the more advanced among the developing countries. Those in the lowest three categories, representing about 46 percent of the world's population, however, are at such a distance from the attainments of the wealthiest countries that significant transfers of resources will be necessary if convergence with the rest of the world is to take place in the foreseeable future.

We have here, then, a division of countries into "worlds" of development that refines the use of GNP per capita as a primary measure of wealth by blending with it numerous other indicators. The OPEC effect also is diminished in importance, and what is often termed the "Third World" appears to be subdivided into four "worlds" in its own right. The distribution of the 94 countries in the three lowest categories at under 36.9 percent of the world standard, is particularly revealing.

Of the major realms, Latin America, Sub-Saharan Africa, and East Asia are the more heterogeneous. The Near East, South Asia, and Southeast Asia are the more homogeneous. The "worlds" of *low, lower,* and *least* development into which these developing countries fall (corresponding to categories of 25.8 - 36.9 percent, 15.9 - 25.8 percent, and less than 15.9 percent of the world standard) display a regional quality that is important both for analytical and policy-making purposes. The "world of low development" clusters in central South America, the Near East, the Caribbean, and southern Africa, with two major outposts in Southeast Asia--Malaysia and the Philippines. The "world of lower development" covers much of the rest of Sub-Saharan Africa and South Asia, but that of "least development" consists of a striking band of poverty running across Sahelian Africa, with a few outliers such as Afghanistan, Nepal, Laos, and Haiti.

Growth Potential Index (GROPOT)

This experimental index (figure 8) for 174 countries applies mainly to the developing countries. Its ingredients refer not only to quality of population, resource endowments, and energy consumption, but also to foreign-aid receipts. *Growth Potential,* as suggested by Papanek, consists of the 22 indicators listed in table 7.[5]

The developed countries fall into the three categories above 65 percent of the world standard, but they are joined in those categories by an otherwise heterogeneous set of developing

---

5. The categories in figure 8 are highly subjective and are not based on a strict quantitative standard.

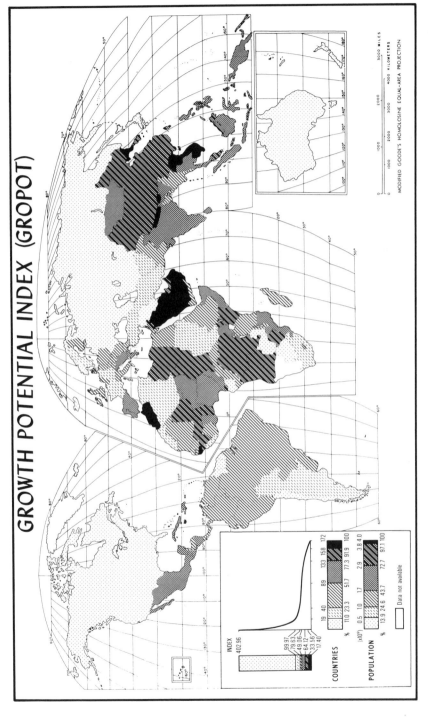

GROWTH POTENTIAL INDEX (GROPOT)

Fig. 8

TABLE 7

GROWTH POTENTIAL VARIABLES

1. Arable Land per Agricultural Population, 1975, Square Kilometers per Thousand People

2. Assistance per Capita, Multilateral Total, 1966-1976

3. Assistance per Capita, DAC and Multilateral Net Concessional, 1969-75

4. Assistance per Capita, DAC and Multilateral Total, 1969-1975

5. Assistance per Capita, U.S.A. Economic, 1962-1976

6. Assistance per Capita, Centrally Planned Economies Total, 1969-1975

7. Calories per Capita per Day

8. Energy Consumption per Capita, Electricity Generated, 1975, Kilowatt-Hours per Person

9. Energy Potential per Capita: Total Energy Potential, 1975, Tons of Coal Equivalents per Capita

10. Energy Potential per Capita, Renewable Energy Potential, mid-1970's, Tons of Coal Equivalents per Person

11. Infant Mortality, 1970, Deaths per Thousand Live Births

12. Literacy Rate, Adult, 1970, Percent

13. Life Expectancy at Birth, 1970

14. Physicians and Dentists per 100,000 People, 1970's

15. Rail Route Length per Capita, Kilometers per 100,000 People

16. Road Network Length per Capita, Kilometers per 100,000 People

17. School Enrollment, Primary, Gross Enrollment Ratio, 1970, Percent

18. School Enrollment, Secondary, Gross Enrollment Ratio, 1970, Percent

19. School Enrollment, Third Level, Gross Enrollment Ratio, 1974, Percent

20. Youthfulness, 1970s, Percent of Population Age 16 or Less, Reciprocal

A high value on the scale means a high growth potential. This required using the reciprocal of the Youthfulness variable when constructing the scale. Countries were dropped from this scale

TABLE 7, *Continued*

if 10 or fewer variables were present. In addition to the above, the following variables were constructed for this scale.

21. Manufacturing Output, 1975, 100,000 US Dollars

22. Exports of Raw Materials, 1973, Millions of US Dollars

countries, including all of Latin America and much of the Near East, excluding Saudi Arabia and the Emirates, which, surprisingly perhaps, rank low on the index despite their petroleum-based prosperity.

The Growth Potential index may be useful as a tentative guide to the future, in particular to the nature of the gap between the richer and the poorer countries. On the basis of this index, it appears that, with effective population control policies and assuming availability of commercial energy, most of Latin America will converge on the developed world in over-all economic development. In contrast, South, Southeast, and East Asia, with the exceptions of Japan, South Korea, and the ROC, would be expected to stabilize or decline relatively. Other realms apparently present too highly variegated futures to permit similar forecasts.

CHAPTER IV
WORLD PATTERNS OF SELECTED VARIABLES

The multivariate analyses described above have shown that many indicators display in varying degree similar distributional patterns at levels of global and regional generalization.   In order to enquire into more specific differences among countries, ten variables have been selected for further mapping and analysis.   The results of this procedure also serve as a partial test of the utility of some of the synthetic indicators.

The variables fall into four groups: product, population, energy, and exports.   These variables have been chosen either because they seem to be analytically discriminatory or because they serve to cast further light on the regionalization problem.   An exception is GNP per capita which, though much criticized, remains the principal means for differentiating among countries in the conventional literature on economic growth and development.

Map Legends
The legends for these variables are organized around a mean value for the variable.   The means are arithmetic, but in the case of GNP per capita, a "world" mean.   The world mean is a mean calculated by weighting each country by its population.   Countries are classified into six categories.   Three categories, each containing equal numbers of countries, are above the mean and three. are below the mean.   The numbers of countries in each group are shown on the top-most horizontal bar.   The lower horizontal bar shows the populations in each category.   The vertical bar shows both the values of the variable and a set of index numbers which are calculated so that the mean is always equal to an index of 100. Similar legends were used in the 1961 *Atlas of Economic Development*. Refer to page 7 of that publication for a more detailed explanation.

Gross National Product per Capita
One of the major problems with Gross National Product per capita as a measure of development is its strong dependence upon

67

exchange rate conversion, a problem mitigated somewhat by measures
that include non-monetary elements. Both Gross Domestic Product and
GNP are denominated in dollars. Kravis et al. argue that "the
lower a country's income, the lower will be the prices of its home
goods and the greater will be the tendency for exchange-rate
conversion to underestimate its real income relative to that of
richer countries."[1] The map legend of GNP per capita (figure 9) thus
has a much steeper and more deeply convex distribution curve than do
the map legends for Factor 1 and the Richdex composite index
(Richdex). Those maps better differentiate among the poorest
countries, most noticeably in sub-Saharan Africa and South Asia.
They also are more discriminating among the resource-rich
underdeveloped countries such as the OPEC group.

To be sure the maps of GNP per capita, Factor 1, and the
Composite index (Richdex) all display a dramatic "North-South"
differentiation. To what extent is this a "political" phenomenon,
as Bauer and Yamey provocatively suggest? "Official foreign aid has
been the unifying characteristic of this huge variegated and utterly
diverse collectivity... Indeed, without foreign aid initiated and
organized in the West, there would be no Third World or South. The
outstanding result of foreign aid is to have created the South."[2]
Whatever the merits of the proposition, and few would probably
accept it, both the developed countries and the lesser developed
countries certainly show considerable variation, a variation which
is illustrated less well by the GNP per capita measure than by the
patterns on the two synthetic scales.

Nonetheless, attempts at regionalization for purposes of
global modeling generally have a political element in that much of
the thought behind the assistance policies of international
institutions follows non-traditional and somewhat eccentric patterns
of political and cultural grouping. Thus, for example, the World
Bank has divided the world into ten regions of which Anglo-America,
Western Europe, Japan, South Africa, Australia and New Zealand, and
the Eastern Bloc constitute the first five.[3] The Bank makes one
region out of Latin America and the Caribbean, and only one of South
and Southeast Asia plus South Korea. China also is one region in

---

1. Irving B. Kravis, Alan Heston, Robert T. Summers, UN International
Comparison Project, Phase II, *International Comparisons of Real Product and
Purchasing Power* (Baltimore: Johns Hopkins University Press, 1978), p. 9. See
also N.S. Ginsburg, "National Resources and Economic Development," *Annals,*
Association of American Geographers, September 1957, p. 199.

2. P.T. Bauer and B.S. Yamey, "East-West/North-South: Peace and
Prosperity," *Commentary,* September 1980, pp. 57-63 quote from p. 58.

3. See IBRD, *The Global Framework,* Staff Working Paper no. 355 (Washington,
September 1979).

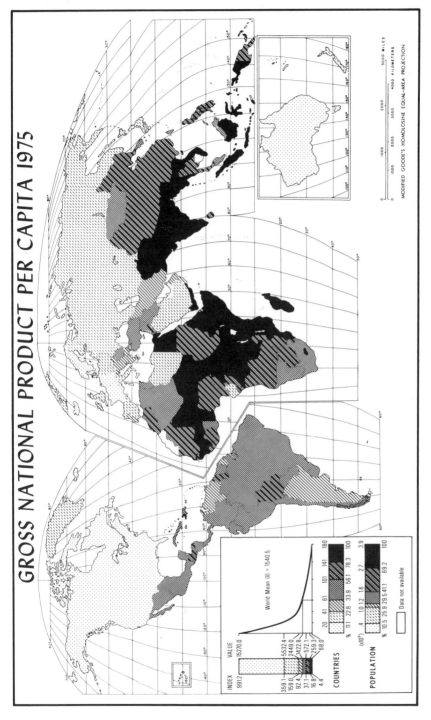

GROSS NATIONAL PRODUCT PER CAPITA 1975

Fig. 9

World Mean (R̄) = 1540.5

MODIFIED GOODE'S HOMOLOSINE EQUAL-AREA PROJECTION

INDEX	VALUE
9912	15270.0
359.1	5532.4
159.0	2449.0
92.4	1422.8
37.1	572.1
16.8	259.3
4.4	68.0

COUNTRIES

%	11.1	22.8	33.9	56.1	78.3	100
	20	41	61	101	141	180

POPULATION

(x10⁹)	4	10	12	16	27	3.9
%	10.5	25.9	29.5	41.1	69.2	100

Data not available

its typology; so too is the Arab World, which, perversely, includes Iran and Afghanistan.

The 120 countries below the world mean on the GNP per capita map, when distributed among the Bank's five lower ranking regions, present almost as much of a differentiation problem as the entire global set. This is why the Bank, and others, turn alternatively to an economic-structural differentiation related to oil. *The World Development Report, 1980,* for example, discusses the energy problem, then the oil-importing developing countries, the oil-exporting developing countries, the capital-surplus oil-exporting countries, and the industrialized countries. All this reflects the lack of easy regionalization associated with both the GNP per capita and the synthetic measures. Capital-surplus oil-exporting but lightly populated developing countries are clearly set off from their neighbors with respect to GNP per capita, and from, for example, populous Indonesia and Nigeria which export oil but not capital and rank low on other indicators of development.

GNP per Capita: Annual Growth, 1960-75

Although it might seem plausible that the poorer countries should have been growing less rapidly economically than the richer, in fact there is little correlation one way or another between national wealth, as measured by GNP per capita, and economic growth. In fact as compared with the global arithmetic mean of 3.17 percent, the industrialized countries averaged 3.78 percent, whereas the developing countries grew at a rate of 3.05 percent. Moreover, as figure 10 indicates, there is surprisingly little regional homogeneity in growth rates in the major world realms. Some generalizations other than that of diversity, however, can be attempted.

For one, the Eastern Bloc countries appear to have been doing better than the predominantly capitalistic Western countries which display greater diversity. The United States, the largest economy of these, falls below the arithmetic mean of 3.17 percent. France and Italy have done considerably better among the developed countries, and the Iberian duo--Spain and Portugal--appear to have done the best of all. Japan, of course, is sui generis among the developed countries. In addition, the data for the so-called Newly Industrializing Countries (NIC)--South Korea, Taiwan, Singapore, Hong Kong, Mexico, and Brazil, inter alia --suggest a gradual convergence between the upper-middle developing countries and the somewhat less wealthy developed ones.

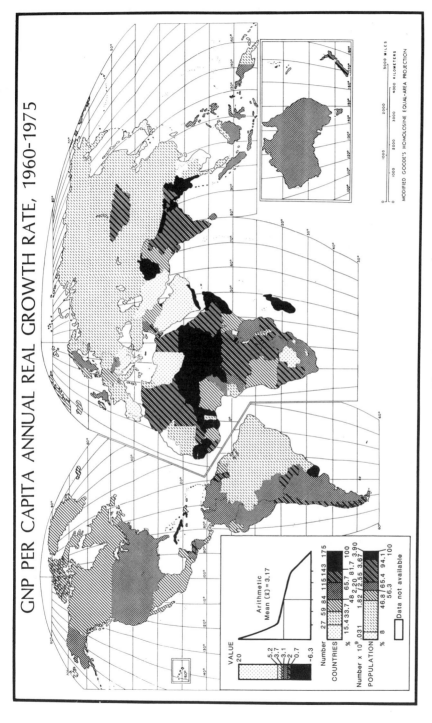

GNP PER CAPITA ANNUAL REAL GROWTH RATE, 1960-1975

Fig. 10

The OPEC effect, or its equivalent in the form of readily available natural resources other than oil, is demonstrated in the high rankings of several of the Middle Eastern countries. The moderately strong performance of a number of other countries such as Malaysia and Thailand is due both to generous resource endowments and to progressive changes in economic structures. On the other hand, with few exceptions, growth in most of Africa south of the Sahara was slight to negligible. South Asia stands out in the same way, and western Latin America, with the exception of Mexico and Ecuador, is similarly laggard, although growth potentials may be high. These clearly are problem regions. In other words, the "poorest of the poor" and most countries ranking just above them appear to have done badly in the global competition, despite the very low levels from which growth has been measured.[4] China stands out as a massive exception among the poor, but the growth rate shown for it of 5.2 percent may be somewhat inflated. Data for the period under examination were scarce and singularly unreliable.[5]

In short, economic growth and developmental status are, on the whole, weakly related; but the gap between the poor and all the rest appears to have been widening, albeit slowly and despite much rhetoric to the contrary. Here lies a major policy dilemma for rich and not-so-rich alike.

Agricultural Production: Annual Growth, 1960-70

In the poorer countries, and a very few of the wealthier, growth of gross national product is highly asociated with agricultural production growth (figure 11), a reflection of the high proportion of the populations of such countries engaged in agricultural occupations. There is a much weaker association between growth of GNP per capita and agricultural production growth, however, due to the intervening factor of differing rates of population growth (see figure 11 following). The correlation between those two variables, that is, GNP per capita and agricultural production growth, is a modest .308. In any event, agricultural production growth in the more highly developed economies (except for Japan) at 1.91 percent per annum lies below the global arithmetic mean of 2.57 percent in the period under

---

4. See L. Berry and R.W. Kates, eds., *Making the Most of the Least* (New York: Holmes & Meier, 1980), especially the Foreword by N.S. Ginsburg, pp. ix-xiv, and chapters 1,2,5,13, and 15 by various authors.

5. The World Bank, however, reports a growth rate for the period of 1960-62 of 5.0 percent. Since rates from 1978 onward were greater than in the preceeding period, the average real rate for 1960-75 presumably must have been nearer 4 percent. See IBRD, *World Deveopment Report* 1984 (New York: Oxford University Press, 1984), p. 218.

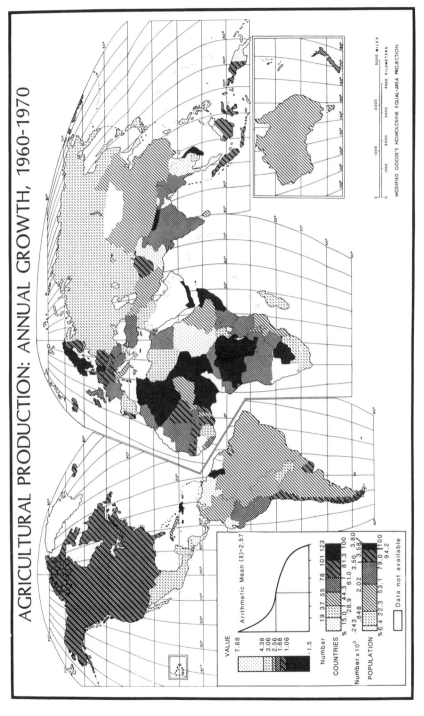

AGRICULTURAL PRODUCTION: ANNUAL GROWTH, 1960-1970

Fig. 11

review. In contrast, the developing countries seem to have done somewhat better. Their aggregate growth rate is calculated at 2.77 percent per annum.

On another dimension, whereas the developed countries display only a narrow range of values, the poorer countries show a very high degree of diversity in their performance. Most of Latin America, for example, lies above the mean; and even some countries in Africa south of the Sahara performed relatively well, albeit from a very low base level--e.g., Sudan, Ivory Coast, and then-stable Uganda. Other African countries, as expected, appear to have performed poorly. In the South Asian realm, only Pakistan did well. In Southeast Asia, Indonesia and Burma did poorly. By comparison, China, even with allowances for an unreliable data base, appears to have done moderately well and better than most.

Gross Domestic Investment: Annual Growth, 1965-73

It is commonly assumed that a close association exists between domestic investment and levels of economic development. The data for some 128 countries, but with some remarkable gaps including the U.S.S.R., indicate that this assumption is incorrect. As shown on figure 12, the wealthier countries in most cases appear to be characterized by investment rates well below the global mean of 5.95 percent, though in Europe during the period under survey, France was just above it. Other important exceptions are those members of the East European bloc whose base levels were somewhat lower than their neighbors to the west and in which high premium was being placed on national investment in heavy manufacturing and in producers' goods as a matter of government policy. In addition, Japan stands out as the capitalist industrial country most likely to become Number One on the basis of aggressive and massive investment. It is not surprising, however, that there should have been a moderately strong correlation (.464 across the 125 countries for which both kinds of data were available) between growth in investment and GNP per capita growth; but, as we already know, there is virtually no relationship between developmental status and economic growth rates. One also might have expected some marked positive relationship between domestic investment growth rates and those of agricultural production; but in fact there was, literally, none (the correlation coefficient between the two was .006). This says something about both the tenuous nature of under-capitalized agricultural growth particularly in the poorer countries, and even more, about the direction of domestic investment--it has not been primarily into agriculture.

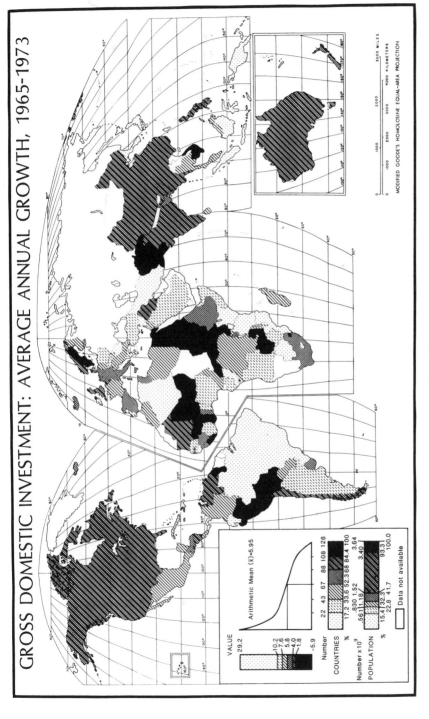

GROSS DOMESTIC INVESTMENT: AVERAGE ANNUAL GROWTH, 1965-1973

Fig. 12

The map shows immense variation in changes in gross domestic investment rates among the countries other than the highly industrialized. Latin America contains examples of all categories in the classification. Brazil stands out as exceptionally investment-oriented, and not from a remarkably low base either; and Argentina and Ecuador did well; but Peru and Bolivia stand out as conspicuously low investors over time. Parenthetically, only in Latin America does there appear to be some positive relationship, though difficult to quantify, between aid flows and investment rates.

In the Near East, most of the oil-producers show high rates of growth, but not all, e.g., Iraq. In Africa south of the Sahara the range of observations was and is difficult to interpret, although, had data been available for Chad, a massive band of low investment almost certainly would have spread across Sahelian Africa and into Egypt as well.

In Asia, low investment growth rates dominated South Asia, except for Sri Lanka. In East Asia, Japan, South Korea, and the ROC (Taiwan) showed high growth rates, as might have been expected, but China seems unusually low perhaps because the years covered also were those of the infamous Cultural Revolution. In the Southeast Asian realm there was substantial variety. South Vietnam and Cambodia fell into the lowest category; and the Philippines, surprisingly, into the next lowest. This might help explain the extraordinarily fragile character of the Philippine economy as it evolved in the early 1980s. However, other countries for which data were available ranked well above the mean, especially Indonesia, but evidence from other sources suggests that the efficiency of Indonesian domestic investment might have been much lower than its rates of growth.

Population: Annual Growth, 1965-75

The global pattern of population growth clearly is associated with that of economic development over-all--the wealthier countries have populations which grow slowly if at all; the poorer have populations with high rates of growth (figure 13). The former have an average rate of growth of .85 percent, as compared with the world mean of 1.91, whereas the developing countries average 2.28 percent. As the map indicates, rates of population growth can be regarded as a virtual surrogate reciprocal of national wealth. Although large populations add to national product, rapid growth rates result in smaller apportionments of that product among them. Thus, it appears

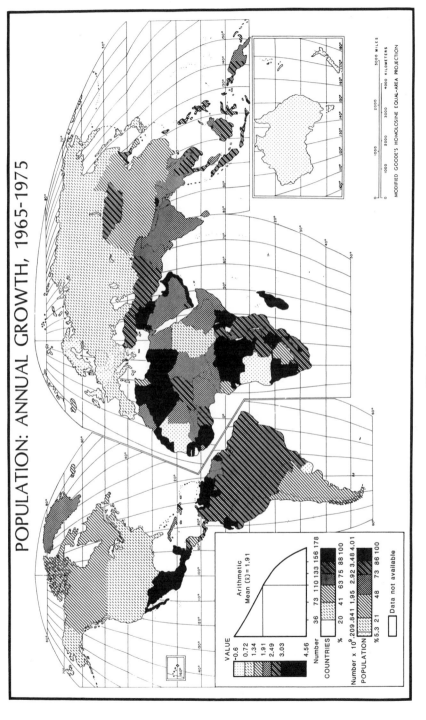

77

POPULATION: ANNUAL GROWTH, 1965-1975

Fig. 13

that, in most cases, rapid population growth may hinder, if not militate against, rapid economic growth.

However, a comparison of this map with that of GNP growth (figure 10) suggests that such a relationship is by no means precise. In fact, population growth is statistically uncorrelated with GNP per capita growth (.058). Nonetheless, this variable is negatively correlated with such indicators of development on a per capita basis as school enrollments, quality of life variables, accessibility, and the Composite Index (figure 7); and it appears unrelated to the nature and volume of international aid flows.

The data for this variable must be viewed with caution, however, given the unreliability of many population data, and also because of more recent information just becoming available. The 1980 Census of India, to be sure, displays the same growth rates as in the period examined here, but data from Indonesia suggest a marked decline in rates of growth. As for China--close to one-fourth of humankind--more precise information must await analysis of the Chinese census which took place in the summer of 1982, but it appears that China is placed here in its proper category.[6] Finally, the data appear to slur over the impact of natural and man-made disasters, as in the Sahel, Cambodia, and Ethiopia during the decade under examination.

Population: Proportion in Cities of 20,000 and More

In the *Atlas of Economic Development,* the textual commentary on the map entitled "Urban Population I" began:

> It is widely believed that the incidence of urbanization in given countries is closely associated with their economic character, that the prosperous will have larger portions of their populations living in cities; the poorer, smaller proportions. To help illumine these assertions this map shows the percentage of national population living in cities of 20,000 or more.[7]

The reader is referred to that earlier map and text which dealt with the difficult problem of defining what is "urban," source materials, and the overall world pattern. Based upon data compiled at International Urban Research at the University of California, Berkeley, for the mid-fifties, they were simply the best then available.

---

6. See Judith Banister, "Chinese Census and the Decade Beyond," in *China: The 80s Era,* ed. N.S. Ginsburg and B.A. Lalor (Boulder: Westview Press, 1984), pp. 173-91.

7. N.S. Ginsburg, *Atlas of Economic Development,* p. 34. See also the map itself on p. 35.

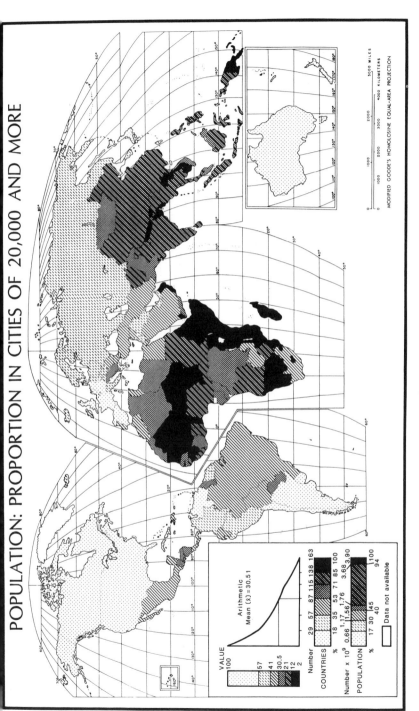

POPULATION: PROPORTION IN CITIES OF 20,000 AND MORE

Fig. 14

The present map (figure 14) is based upon materials from the mid-70s for 163 countries compiled by Richard Forstall of the U.S. Bureau of the Census.[8] These data probably are the best there now are, and the map based upon them is the only one of its kind. It should be noted that data, of variable quality to be sure, can be found elsewhere for most cities and/or metro areas of 100,000 and more in the world, but the 100,000 cut-off does not faithfully represent a practical distinction between what is urban and what is not. The 20,000 figure, we believe, does, but that is a topic for another dissertation.

In any event, the distributions for the mid-1970s are remarkably similar to those of twenty years earlier. Wealth and high proportions of urban populations go together, at least for the developed countries. On the other hand, the correlation between GNP per capita and urban population is not high, less than .300. The reason for this is seen in the enormous range of values for this variable among the lesser developed countries. Although most countries in sub-Saharan Africa and the three Asian realms (other than Japan and the other Sinitic outliers)--East Asia, Southeast Asia, and South Asia--rank well below the world mean of 30.51 percent, the countries of the Middle East (including the Maghreb) for the most part lie somewhat above the mean. Even more important are the high ranks of most Latin American countries.[9]

It is important to note, however, that whereas the world mean in the 1950s was 21.6 percent, the value of the mean in the 1970s had risen to 30.5 percent. Moreover, that increase reflects larger relative increases in the poorer countries than in the richer. In fact, nearly two-thirds of the world's urban population is now believed to reside in the developing countries and nearly half in Asian countries alone. The implications for the future are immense.

Energy Consumption: Commercial as Percent of Gross

Energy consumption, and particularly commercial energy consumption, have been regarded as virtual correlates of modernization and industrialization. Indeed, the data on energy

---

8. These data arrived too late to be included in the computerized dataset, so they are not mentioned elsewhere in the text or in the lists of variables. The data are unusual and of such importance that we decided to map them and include them in the text.

9. This then raises the question of functional and structural differences among cities in developed countries as contrasted with those elsewhere. Much of the literature on this subject suggests that these differences are substantial, but the topic must be treated elsewhere. The problem is discussed, for example, in J. Osborn and A. Atmodirono, *Services and Development in Five Indonesian Cities* (Bandung Institute of Technology, 1974).

ENERGY CONSUMPTION: COMMERCIAL AS PER CENT OF GROSS

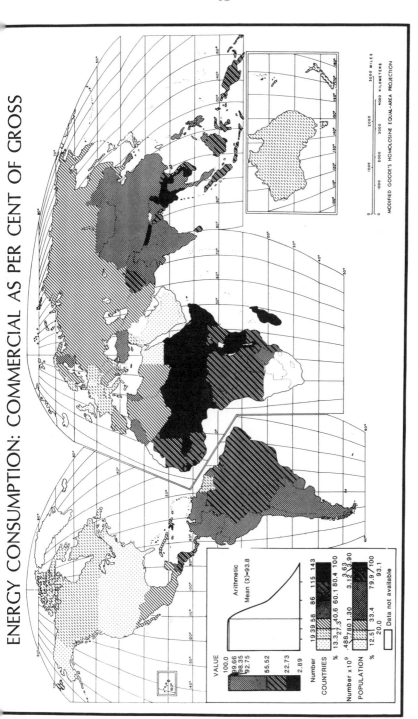

Fig. 15

consumption, when mapped, reveal global patterns similar to those of GNP per capita, of Factor 1, and of the Composite index. However, recent data indicate that the correlation coefficient between commercial energy consumption per capita and GNP per capita, for example, is only a moderately high .612, indicating some variation among the developing countries. Rather than presenting those materials cartographically here, a somewhat more sensitive measure has been chosen for presentation on figure 15, the proportion of gross energy consumption that is of commercial origin, for which data were available in 1975 for 143 countries.

The patterns shown on figure 15 and its legend are quite revealing. All of the developed countries, excepting Finland, lie above the world mean of 93.8 percent. So also do the oil-producing Middle Eastern countries. Surprisingly perhaps, several of the other OPEC countries lie below the mean, e.g., Venezuela, Nigeria, and Indonesia, but these all have large rural-based, poor populations. All of the poorer countries, even several of those in the newly industrializing category, depend substantially on non-commercial sources of energy, and the legend curve drops off sharply below the mean for sixty percent of the countries surveyed.

Alternative sources of energy to the conventional commercial ones are wood and waste, e.g., dung, though it should be noted that wood may be marketed and therefore is "commercial" in the strict sense of the term. The comparatively high dependency upon these alternative energy sources is marked by massive deforestation with consequent problems of soil erosion and environmental degradation, as in Brazil, where the non-commercial enery consumption is some 46 percent of the total, and by declines in soil fertility through the burning of animal wastes as fuel, as in India where the percentage is near 30 percent. In most developing countries outside of the Middle East and excepting Mexico, therefore, means must be found for energy generation that will be passive with regard to the environment and independent of outside supplies of commercial fuels for which precious foreign exchange may not be available. Biogas generators, as being developed in China and India, are one such means. Elsewhere it has been argued that, in the humid tropics at least, hydroelectric energy may be the answer,[10] but the problem of making electricity available in rural areas at costs compatible with deep rural poverty (figure 8) may prove to be insurmountable.

___

10 . N.S. Ginsburg, "Energy and the Poorer Countries: The Context of a Strategy," in *Energy Technology and Global Policy,* ed. S.A. Saltzman (Santa Barbara: Clio Books, 1977), pp. 195-203.

Energy Consumption: Commercial per Capita Growth, 1965-1975

Generalizations about the global pattern of commercial energy consumption per capita growth are, of necessity, very broad and encompass many exceptions (figure 16). Two factors, population size and the initial or base point rate of commercial energy consumption, have combined in varying degree to mould the pattern. Countries with small populations and a very low initial consumption rate tend to have made the most rapid increases on a percentage basis. Conversely, those countries with higher base point rates and/or large, dense populations, tend to have a rate of increase which is below the world mean. However, the fact that the United Kingdom and Kampuchea fall into the same category indicates the potential significance of the exceptions to those two tendencies, that is, similar rates of growth may be symptomatic of totally different situations in different countries. In any event, in most developed countries growth rates have been low, as has been the case for economic growth in general, with the conspicuous exception of Japan.

Interestingly, the division between northern and southern Western Europe is more pronounced than that between the so-called East and West. Northern European countries tend to have low rates of increase, while those of southern Europe rank above the mean. The U.S.S.R. is ranked in category three above the mean, but in fact it is only slightly above it.

Among the poorer countries, most of the oil-rich countries have shown considerable improvement. In Latin America, the Andean states, other than Ecuador and Bolivia, have done poorly, but so has oil-rich Venezuela. In Sub-Saharan Africa the pattern is highly variegated, but initially impressive gains in some of the poorest countries almost certainly reflect an extremely low base point. In Asia, some of the smaller countries show high rates of increase; so does Indonesia, which is encouraging, but its base point was very low. In general, the larger countries in Asia have done poorly. As a result, one finds that only 35 percent of the populations in the 151 countries treated resides in states which have increased their commercial energy consumption faster than the average.

Increases in energy consumption, like those in GNP per capita, bear little relation to various measures of current wealth. Moreover, although rapid increases suggest the possibility for improvement in the developmental status of the poorer countries, the significance of such changes will vary with almost every individual case. Countries with current high levels of consumption, e.g., Spain, indeed may be assumed to have high potentials for rapid

84

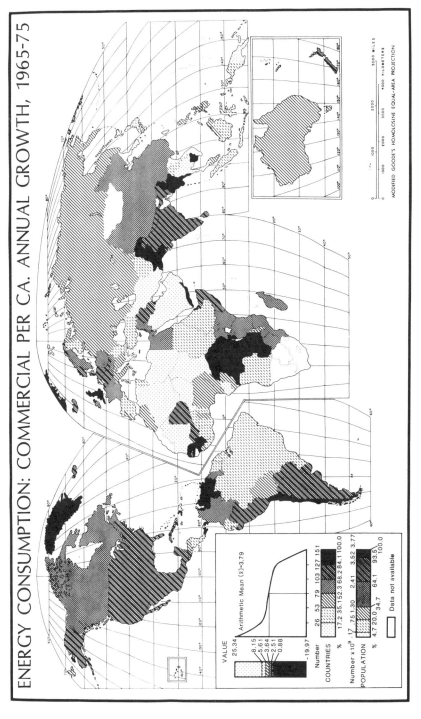

ENERGY CONSUMPTION: COMMERCIAL PER CA. ANNUAL GROWTH, 1965-75

Fig. 16

economic growth, but such countries are few and do not dominate the
high-ranking array. On the contrary, the top ranks are occupied
mostly by countries with small absolute increases based upon
extremely low initial rates of per capita consumption.

Exports: As Proportion of GNP

Exports, as one part of over-all international trade turnover,
are a measure, though by no means an exclusive one, of the
involvement of a given national economy with others.[11] Exports, as
part of that turnover, are particularly important since they are
largely the means whereby countries generate foreign exchange, pay
for imports, and in general balance their national accounts. For
the developing countries, the export trade is especially important
for future development.

Figure 17 demonstrates that very large economies, whether
richer or poorer, have merchandise export trades which are well
below the world mean of 12.17 percent of gross national product.
This is true of the United States and the USSR, for example; it also
is true of China, India, Brazil, and Mexico. Most other developed
countries show exports as high proportions of GNP, as do the newly
industrializing countries; so do most smaller, poorer countries
whether rich in a resources like petroleum or otherwise dependent on
other raw-material exports. The latter point holds also for
Indonesia as a major oil and estate-products exporter.
Interestingly, Japan also shows a lesser dependency on exports than
might have been anticipated. To be sure, all highly developed
countries export services (and often capital) which do not show up
in these data.

There is in fact little relationship between this variable and
the usual development indicators such as GNP per capita, life
expectancy, etc. This might suggest that exports per se need not
lead in economic development, but that proposition probably holds
only for the largest economies, especially those relatively well
developed, and even then will vary among them. For the smaller,
poorer countries exports are crucial. For those without major
natural resources or industrial products suitable for export,
dependence upon the largesse of the wealthier countries may well be
unavoidable.

---

11. Since exports are not part of the Gross National Product, their value
under certain circumstances may exceed GNP itself, as in the cases of Hong Kong
and Singapore, for example, the entrepot trades of which are very large. Of
course, the values of services associated with those trades are incorporated into
the GNP.

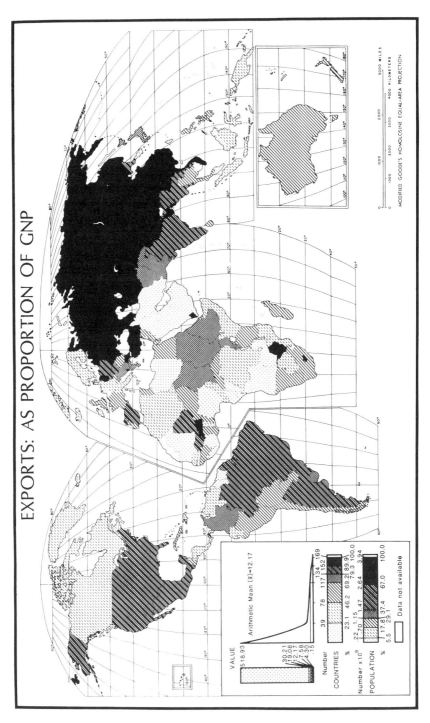

EXPORTS: AS PROPORTION OF GNP

Fig. 17

Exports: Concentration Index

The global pattern revealed by the concentration-of-exports index map (figure 18) is an intricate mosaic about which it is difficult to generalize.[12] No one factor sufficiently explains the rank-order of countries with respect to this variable. It is true that the countries generally accepted as "developed" all fall into the lowest categories where diversity in exports is high. The index for 26 economies (154 or less) is only a little more than a third of the global mean (433). However, no such generalization can be made about the so-called lesser developed countries which exhibit considerable variation in their index values. Several of the poorest and least developed countries rank well below the world mean; others have a very high concentration index.

The intraregional patterns and the contrast between them are, perhaps, the most notable aspect of this map. South and East Asian countries, including the Indian and Chinese giants as well as the newly industrializing countries, e.g., S. Korea and Taiwan (ROC), tend to have a low index value, falling almost exclusively in categories one and two. Southeast Asia exhibits considerably more variation, with Kampuchea, South Vietnam, and Brunei ranking well above the world mean of 432.83, but others falling well below it even when dependence on a few exports is high.

The regional patterns of the Near East, sub-Saharan Africa, and Latin America are similar in both their complexity and their range of values. In all three of these realms, the countries range from the highest to the lowest categories. Mexico, Lebanon, and Southern Africa have considerably lower concentration indices than their neighbors. The OPEC type of effect, however, is clear in countries like Iran, Iraq, Saudi Arabia, Libya, Nigeria, and Venezuela where oil dominates exports, and in Bolivia, Chile, Zaire, Chad, et al., where exports of one or more other natural products dominate.

North America and Eastern and Western Europe are uniformly well below the world mean. The countries of these three realms dominate the lowest rankings and thus have more diversified export bases than the countries of other realms.

---

12. This index, calculated by Daniel Dzurek, is based on a normalized Hirschman index defined at the three-digit SITC level. The data are from the *United Nations Handbook of Trade and Development Statistics* (Geneva, 1976). In that source, the data are normalized between 0 and 1. In this volume, the data were multiplied by 1000 so that the possible range is between 0 and 1000. The original index is from Albert O. Hirschman, *National Power and the Structure of Foreign Trade* (Berkeley: University of California Press, 1945), see appendix A.

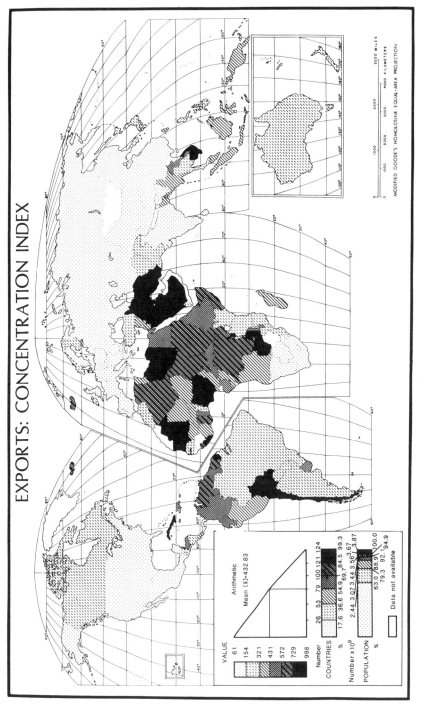

EXPORTS: CONCENTRATION INDEX

Fig. 18

This map then bears little resemblance to the map of Gross National Product per capita (figure 9). Afghanistan and Haiti, for example, are among the poorest and least developed countries of the world; yet they apparently have a relatively diverse export base. This map does not indicate the volume or value of exports, however, and it cannot, therefore, indicate their wealth-producing potential.

It is, nonetheless, a good indicator of the complexity and diversity of national income-generating production functions among the less developed countries. There is a significant correlation, too, between this concentration index and export dependency on raw materials (.527 for 123 countries), which supports the linking of these two variables in much of the developmental literature. Countries which are dependent upon a few export products are much more vulnerable to production and market fluctuations. Whether the major export product is oil, rice, tin, or sugar, a relatively small shift in supply or demand can cause a serious imbalance in any national economy with a high concentration of exports index.

Exports: Proportion to Developed Countries

The received wisdom has it that the poorer countries trade largely with the developed ones and that their dependency upon the "first world" of development is extremely high. This map (figure 19) shows through its variegated patterns how questionable this proposition might be. To be sure, the world mean for this variable is just under 63 percent, a high percentage, and 60 percent of the countries of the world lie well above that mean, but there also is considerable variety among them. Another widely held belief is that the developed countries trade with each other primarily, but in fact that proposition also needs qualification. In any event regional differences abound.

The clearest situation appears to be that of Africa south of the Sahara, most of the countries of which are heavily dependent on export trades with the developed countries. Most others lie close to the world mean. Exceptions in eastern Africa may reflect dependence on a neighbor for transit, as in the cases of Sudan and Zambia, or very low export volumes, as in the case of Ethiopia. In East Asia, China ranks low on the scale, and in South Asia so do all the countries in the realm. Size again appears to matter there. Much variety exists in Southeast Asia, though Malaysia probably has a higher real value than shown on this index since much of its export trade flows through Singapore. The Philippines in contrast are an exemplar of dependency. In Asia, too, Japan's low value

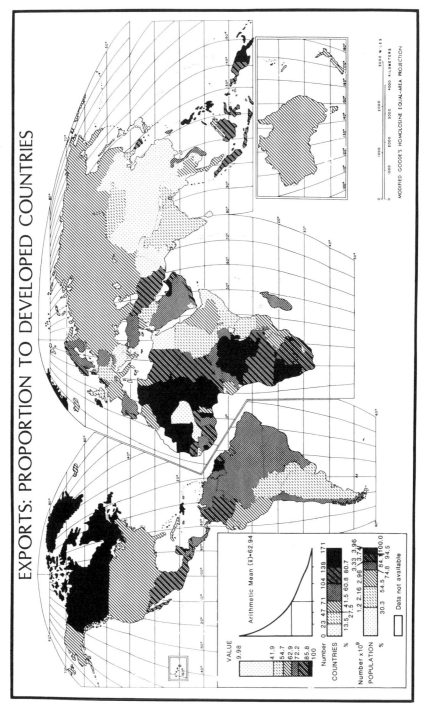

EXPORTS: PROPORTION TO DEVELOPED COUNTRIES

Fig. 19

might seem surprising, but even as of the mid-70s and despite the
role of the United States as Japan's premier market, Japan's trade
was with the entire world, and nearly a third of it was to other
Asian countries alone. In Latin America, most countries show high
or middling values, but several--Argentina, Uruguay, Bolivia, and
Paraguay--have lower values, near fifty per cent dependency. Of
these, Bolivia seems the most remarkable, but much of its trade as a
land-locked state is transit trade.

The developed countries, on the other hand, rank relatively
close to the world mean. Some, like Eire, the Scandinavian
countries, and the Low Countries obviously interact mostly with
their neighbors, but the rest seem to lie near or well below the
mean, as might be expected of global economies which, in the
mid-70s, were heavily dependent on imports of petroleum from the
OPEC countries and exported goods as well as capital in return. The
United States and the Soviet Union also fall near the mean, but
Canada lies far above it, reflecting its deep involvement with the
American economy.

How might these patterns appear a decade hence? Almost
certainly the export relationships between most African countries
and the more highly developed ones will be intensified. To a lesser
degree that trend might also apply to Latin American countries as
well, but as these countries continue to industrialize, e.g.,
Brazil, exports to their neighbors and to other lesser developed
countries also can be expected to expand. This most probably will
be true of India and China also, but forecasting must remain an
uncertain business. What the map cannot show, however, is the
extent to which the industrialized countries are independent of the
poor, despite much received wisdom to the contrary. If that degree
of independence is high, as we believe it to be, then the futures of
many poor countries, heavily dependent on exports to the rich, are
even more suspect than current information suggests.

CHAPTER V
ENVOI

This report is another stage in a continuing investigation by the authors in the geography of economic development. It is an exploration of various ways to approach the regionalization of economic development and the problem of discriminating among countries which normally fall within similar categories of wealth, as measured by Gross National Product per capita and other, perhaps more sophisticated, measures. The results appear to be modest, but it is hoped that they will stimulate further enquiry by those interested in developing a more useful, and a more relevant, description of the world economic order.[1]

The proposition that Gross National Product per capita as a measure of comparative national development may conceal more than would be desired for purposes of analysis and policy formulation has been tested and seems to have been confirmed. On the other hand, its continuing high value as a crude measure of wealth among countries has been substantiated by the mapping and comparison of its global patterns with those of other measures. Factor 1, the "modernization" measure, for example, is highly correlated with GNP per capita. However, when that measure is analyzed cartographically, the OPEC effect is significantly reduced, both Latin America and Africa are shown to be highly differentiated, and the conventional rich nation/poor nation dichotomy is seen to be significantly blurred.

Furthermore, the fact that it has been possible to define statistically and map the bivariate patterns which lie within Factor 1 is important. Gross National Product per capita loads higher on the component called the Transportation scale, but not so high as to replace it; and it also loads high on the alternative dimension. Therefore, plotting that subfactor against the other sub-factor (the Social Services scale) has significant analytical utility in

---

1. Even these modest results required the accumulation of large quantities of information. Readers interested in obtaining copies of some of the data compiled, which the constraints of space forbade publication of in this volume, may write the senior author for print outs.

relating the wealth of 143 nations to their apparent capacities to distribute it.

The patterns formed by mapping the special indices at a "World Standard" even more dramatically describe the complex distributions of developmental structures on broadly defined continua. This is true especially of the Composite index (RICHDEX).

The special indices also are useful in identifying the developmental attributes of any one country and in comparing its developmental distance from successful achievement with those of other countries.[2] Sri Lanka, for example, scores 29 on the Composite (RICHDEX) scale, relatively high for a poor country, lower on indices of accessibility and urbanization, and higher on measures of quality of life, industrial economy, and non-diversified trade as a trading nation. Still, its distance from maximum growth potential (GROPOT) is long. These positions of Sri Lanka on the several special scales reflect a distinctive mix of attributes which set Sri Lanka off from both its geographical and statistical neighbors on conventional GNP per capita scale. Moreover, Sri Lanka does relatively well on specific measures of growth--e.g., agricultural development, gross domestic investment and exports which are seasonably diversified. In short, this is no textbook underdeveloped country.

Country profiles apart, global patterns of development as revealed by mapping the "special-index" data are of interest specifically as they help identify (a) the several real "worlds" of development, (b) the regionalization of that development, and (c) exceptional countries. On the basis of available development indicators, it appears that the commonly recognized three "worlds" of development, even the so-called "Third World," have little value for the comprehension of economic development patterns. What appears to exist rather is a large group of industrialized countries, including those of both the Western and Eastern Blocs and their neighbors mainly in Mediterranean Europe, which can be considered the "developed world" (at above 50 percent of the RICHDEX composite world standard). The rest of the world, on the other hand, can be divided into at least four "worlds" or "sub-worlds" in descending order of wealth and intensification of poverty. Although all are relatively poor in terms of distance from the "developed world" nations, they may be differentiated, as described earlier (pp. 59ff.) into the "more developed" poor and the "low," "lower," and

---

2. Only two of these special indices have been mapped and discussed in this article--RICHDEX and GROPOT. See pp. 59ff. above.

"least developed" among them. In this taxonomy the larger OPEC members join countries having little or no positive OPEC effect, but which have been developing substantially nevertheless. Moreover, China is set apart clearly from the Indian sub-continent, and countries at the extremes of human misery are identified with remarkable precision. In addition one of the important qualities of this taxonomy, which may, however, subject it to some criticism, is its detachment from ideology. The "socialism vs. capitalism" dyad is regarded as secondary and derivative, not primary and causal-explanatory.

Beyond these admittedly somewhat static patterns of relative poverty and wealth are other patterns based upon dynamic spatial developmental structures. Thus, we may speak of "worlds" not only of national wealth but also of potentials. In brief, the presently economically powerful states are generally not those displaying high economic growth rates, with the exception of Japan--quite the reverse. In addition, though population size is closely associated with national product and therefore gross wealth, population densities often are not, and high population growth rates appear to move against the grain of evolving developmental structures in most of the world.

These limited generalizations may be complemented by several others which are strongly suggested by the data and the analyses of them, but which are far more speculative and will require careful and considered further consideration.

For example, it appears that the developmental gap between the "first world" composed of the industrialized countries, and perhaps the newly industrializing ones, and the poorest countries, that is, those in the two lowest categories of the several sub-worlds proposed above, is increasing. This appears to be the case despite the slowing in growth of the wealthier countries in the first category. It follows that those poorer countries especially in parts of Sub-Saharan Africa, will become ever more dependent on international aid flows from the richer unless they were to lose their identities by merging with neighboring countries of higher growth potentials, if such exist. Even if they do, such mergers would be unlikely in the contemporary world, let alone the even more unlikely outcome of merger with some distant, wealthier country in a contemporary version of a benign colonialism. A hallmark of the global political order is an intense--and some might say virulent--form of Nineteenth Century nationalism which places sovereignty and territoriality at the forefront of national

political objectives. The alternative of autarky is not available
to most such countries since their resource endowments usually are
meagre; they also commonly are fragmented ethnically and lack a
well-defined common purpose other than survival. To be sure, this
dour scenario cannot be fully supported without further time-series
data and more intensive examination of individual countries, but the
handwriting appears to be on the wall.

Second, the poorer countries which constitute the sub-worlds
nearer the "Third World" appear to be moving hesitatingly and with
great difficulty toward comprehensive and self-sustaining
development. To be sure, a China, an India, and even a Brazil, or a
Thailand and a Malaysia, are not likely to reach the per capita
levels of income that are associated with the industrialized states,
at least within the foreseeable future, but they already appear to
be on a course largely free from dependency on the largesse of the
developed worlds. In all cases, much will depend on a slowing of
over-all population growth, so that short-term gains in production
will not simply be consumed by a rapidly growing population. In
this connection, uncertainty about futures in much of Latin America
will be more problematic than for those in most of South Asia,
Southeast Asia, and East Asia.

Third, the contrasts between the inter-tropical regions and
the rest of the world continue to be as sharp as ever. It is a
remarkable fact that, of all the developing countries, only China is
in the middle latitudes and therefore can benefit directly from the
vast investments in the agricultural sciences which have greatly
increased agricultural productivity, both per unit area and per unit
labor input, over the past century. Such agricultural research as
went into the low-latitude regions, until about 20 years ago, was
associated largely with a limited array of estate or plantation
crops. The more recent research and development done at Los Banos
in the Philippines, which has led to the so-called "Green
Revolution" in low-latitude rice production, represents a major
breakthrough no doubt, but the emphasis on high-yielding paddy
varieties that require large inputs of fertilizers and water control
measures has contributed to various problems which have mitigated,
in part, the favorable outcomes of the innovation. What is needed
is research into higher yielding, less photoperiodically sensitive
varieties of paddy that do not require such massive inputs.
Moreover, apart from rice, relatively little research has been
directed so far toward other low-latitude and inter-tropical crops
such as cassava and yams which constitute a major basis for much

low-latitude agriculture particularly in sub-Sahara Africa. Clearly
a mobilization of scientific effort in this direction is a sine qua
non for the alleviation of rural poverty in many of the poorer
countries.[3]

Fourth, the regionalization of underdevelopment, which
appeared so strong 30 years ago,[4] that is, the delineation of
regional types of development and developmental problems, appears to
have become rather less well defined in the decades since. There is
a paradox here in that it has become common in recent years to
espouse regional solutions to developmental problems, as illustrated
by the recently completed Convention on the Law of the Sea (1982)
which enshrines such regional solutions in constitutional terms.
Moreover, one has difficulty in identifying international regional
associations which are capable of dealing with developmental
problems on a regional basis. Even the most successful of these,
the Association of Southeast Asian Nations (ASEAN), falls well short
of what might be required, and other regional associations show
little sign of becoming effective developmental institutions.[5]

Finally the unique characteristics of individual countries, so
well identified on many of the maps presented in this study, seem to
be becoming of increasing importance in devising strategies for
successful developmental programs. To be sure, one cannot even
begin to think about such programs without recourse to
generalizations that extend far beyond national territorial
boundaries. Nonetheless, the importance of the idiosyncratic is
underscored by the distinctive mixes of characteristics that are
associated with individual countries. Effective policy formulation
must be based upon the careful and intensive study of the human
geographies, if you will, of given polities and societies. What is
good--and right--for Sri Lanka might not be appropriate for Tanzania
or Peru. This is a conclusion arrived at somewhat reluctantly. It
would be so much easier to simply argue in general terms for
investment in human resources, for example; but one is confronted
immediately with the question: what sort of investment, for what,
and for what purposes? Then the idiosyncratic comes to the fore.

---

3. These problems and strategies are reported on in the *World Food and
Nutrition Study: Potential Contributions of Research* (Washington: National
Research Council, Commission on International Relations, 1977).

4. See B.J.L. Berry, pp. 116-7, in the *Atlas of Economic Development,* 1961.

5. For a discussion of regional approaches to problems in the Southeast
Asian seas, see George Kent, "Regional Approaches to Meeting National Marine
Interests," *Contemporary Southeast Asia* 5, no. 1 (June 1983), pp. 30-94.

In keeping with these propositions one must be aware of the value of different analytical approaches. The diversity of the principal components of the existing world order--the nation states--can be new to no thoughtful observer. Comprehending that diversity through the systematic examination of commonalities and differences is a continuing challenge to scholars and policy-makers alike. Perhaps this monograph will act as one small step toward attaining that goal.

MULTIVARIATE METHODOLOGY IN CROSS-NATIONAL RESEARCH

MULTIVARIATE METHODOLOGY IN CROSS-NATIONAL RESEARCH

Grant Blank

Introduction: The Goal of Methodology

How material welfare is measured and analyzed plays a crucial role in the conclusions that can be reached. Although a general· exploration of how methodology influences substantive conclusions is beyond the scope of this chapter, it is important to indicate some of the ways in which this monograph was influenced by the methods chosen. That is, in fact, the underlying theme of this chapter. As the reader will notice, there is an immense variety of decisions to be made and a great number of details to consider. The goal of this chapter is to make the reasons for those decisions explicit and to point out why and how the details were handled.

The methods used are always choices from a variety of possibilities. This leads to the question, what are the criteria for choosing certain methods and rejecting others? The substantive assumptions about the nature and scope of this study had the strongest influence on the choice of methods. The fundamental assumption was that much work needed to be done to refine understanding of the global dimensions of wealth and poverty. Thus this study was designed to pay attention to world patterns on three levels: globally, at the level of individual countries, and at the level of groups of countries (especially regions).

Further, the methods chosen must focus on the potentially theoretically interesting aspects of the data: differences among countries on different dimensions and attempts to find regularities and common categories. Conversely, they should limit the effects of those aspects of the data that were of no theoretical importance. An example of a theoretically uninteresting aspect of the data is the units of measurement--say, dollars or thousands of dollars. This is irrelevant substantively but in some cases can have a major impact on statistical procedures.

Finally, the methods must reflect, as closely as possible, understanding of the nature of material welfare as an empirical phenomenon. They must be, in this sense, appropriate.

Realizing this, we have based our choice of methods on three assumptions:

1. Rather than attempting to confirm one or more existing theories, our research was exploratory. Thus, our analyses required techniques which assumed as little as possible about the nature and structure of the data. In other words, our methodology needed to be open maximally to the patterns revealed by the data.

2. The fundamental characteristic of material welfare is that it is complex and multidimensional. A variety of indicators is required to capture the full range of reality. Appropriate data analytic techniques are those able to handle these multiple dimensions.

3. These multidimensional data cannot be analyzed by any single technique, if only because we are not sure that there is a unique "best" technique. Each technique has particular strengths and weaknesses. A variety of complementary techniques is required.

To these ends we employed the statistical techniques discussed below. The chapter is divided into four parts. It begins with a discussion of possibly the most far-reaching decision that was made: What variables would be included in the analysis. It is a companion to the second part which discusses which countries were included in the dataset. This part pays particular attention to the concern with avoiding the biases caused by small, unrepresentative samples and the need to achieve coverage of as many countries as possible. Part three examines the major problems that we faced in preparing the data for analysis. The final part ends the chapter with a discussion of the techniques employed, the justification for those techniques, and the problems encountered.

Choice of Variables

Certain fundamental decisions concern the selection of variables to be examined. Two major criteria were used to select the variables for analysis. First, some variables were chosen because theories of economic growth and development indicated that they were important. This category includes major variables like Gross National Product, literacy, urban population, steel consumption, and others. The text of the 1961 *Atlas of Economic Development* indicates why many of these variables are important.[1]

---

1. Norton Ginsburg, *Atlas of Economic Development* (Chicago: University of Chicago Press, 1961).

Second, other variables were chosen because they were surrogates for some in the first category and could be used to cross-check results. An example of this is the different energy consumption indices.

A careful reader will have noticed that many potentially interesting variables were not included. The choice of variables reflects several criteria. First, as the title indicates, the focus of this study is on patterns of economic development. Variables reflecting other concerns, for example, political change or institutional and social structure, were not examined, although the issues that these variables are capable of addressing are of major importance in other contexts. Second, a major priority of this study was to cover as many nations as possible. If data could not be found for at least 100 nations the variable was not included. Third, several interesting variables were omitted because, even though they had been included in the 1961 *Atlas,* they were no longer deemed significant. Among these are two major variables; Commercial Energy Consumption (map 35 in the 1961 *Atlas)* and the Proportion of Waterpower Potential Developed (map 40 in the 1961 *Atlas).*

Coverage

The extent to which data were available for many countries, that is, the "coverage" of the variable, was an important issue in this study. Since the coverage issue is often misunderstood, it deserves extended discussion.

Many studies of economic development are limited to a subset of countries. One commonly used criterion for exclusion is country population. Typically countries of less than one million people are excluded.[2] Another is difficulty in obtaining data (e.g., North Korea or the People's Republic of China). A third reason for exclusion is the fact that the country's political system makes price and production information non-comparable with the rest of the world. For this reason, socialist countries are often excluded from comparative studies of development.[3] Finally, some studies focus on only a single region, usually Latin America or Africa, since those are the only regions with enough countries to avoid sample size

---

[2]. Cf. Gustav Papanek, "Aid, Foreign Private Investment, Savings and Growth in Less Developed Countries," *Journal of Political Economy* 81 (1973): 120-130 or Volker Bornschier, Christopher Chase-Dunn, and Richard Rubinson, "Cross-National Evidence of the Effects of Foreign Investment and Aid on Economic Growth and Inequality: A Survey of Findings and a Reanalysis," *American Journal of Sociology* 84 (1978): 651-683.

[3]. E.g., Richard Rubinson, "The World-Economy and the Distribution of Income within States: A Cross-National Study," *American Sociological Review* 41 (1976): 638-59 and Richard Rubinson, "Dependence, Government Revenue, and Economic Growth, 1955-1970: A Cross-National Analysis," *Studies in Comparative International Development* 12 (1977): 3-28.

problems.[4]

As a result of these practices many studies base their conclusions on an extremely small number of cases. The reasons for exclusion often are readily understandable, and they were considered in this study as well. We experienced first-hand the difficulty of gathering data on many countries; and prices really do not have the same meaning in a command economy as they do in market economies. Despite these problems, we believe that there are three overriding reasons for choosing to gather data on every possible country. First, when sample size is small and the range of variation has been limited by the researcher, the results of the study are open to question on statistical grounds. The results may be a statistical artifact due to errors caused by sample restrictions. Regardless of the true pattern of relationships the researcher is investigating, it is still possible that any limited subset of countries may show a different relationship than the entire set of countries.[5]

Second, some of the reasons listed above for excluding countries are not theoretically grounded. For example, there is no theoretical reason to believe that the economies of countries with less than one million people should grow and develop differently than the economies of larger countries. Similarly, even though prices are different in Socialist countries, and gross national product (or other indicators based on price) is therefore not exactly the same thing in a Socialist economy as in a capitalist economy, the concept of an aggregate level of economic activity is equally meaningful and very important. Merely because a variable cannot be *simply* measured or *easily* compared across countries does not mean that the concept behind the variable is meaningless. Nor does it mean that we should not attempt to carry out the measurement (with, of course, appropriate caution and caveats). In short, if the variable was significant, we tried to find data to measure it but remained aware of the potential problems involved.

The third reason for seeking maximum coverage is grounded in the task of this research. The goal is an investigation into the *global* characteristics of economic growth. Secondary interest is given to international policy intended to foster that growth. Countries that are small or closed to the international community or are Socialist also have economies that will grow (or not grow, as

---

4. E.g., Keith Griffin and J. L. Enos, "Foreign Assistance: Objectives and Consequences," *Economic Development and Cultural Change* 18 (1970): 313-27 or Patrick McGowan and Dale Smith, "Economic Dependency in Black Africa: A Causal Analysis of Competing Theories," *International Organization* 32 (1978): 179-235.

5. Bornschier, Chase-Dunn, and Rubinson, "Cross-National Evidence of the Effects of Foreign Investment and Aid on Economic Growth," p. 670.

the case may be) and develop. They cannot be ignored if we intend
to examine the broad sweep of economic growth across the world.

Data Preparation

Preparation of the data prior to analysis is a time-consuming
and error-prone task. The crucial goal is to put the data in such a
form that statistical and methodological artifacts will be
eliminated (or at least minimized). Since different methods are
sensitive to different problems, data have to be prepared with a
clear eye toward which analytic techniques will be used.

This research confronted many of the typical problems of
multivariate social science research. This section discusses the
three most crucial problems. (1) How to minimize the distortions
caused by differences in the scale of measurement of our variables.
(2) What to do about the skewed distributions of most variables.
(3) Ways to handle the endemic problem of missing data. Our
solutions to these problems are applicable to almost all of the
research using cross-national data. In fact, these difficulties
plague most multivariate research, and our solutions are generally
applicable across a wide variety of data and research topics.

In other research, data preparation will involve several
additional problems. One common problem would be the need for much
more care and cross-checking during data collection. This would be
typical of survey research or other studies relying on responses
from subjects for their primary data. We had the advantage of being
able to draw primarily on published data from various international
agencies rather than conducting our own data collection. A second
problem involves database design. Our data were gathered for a
relatively simple, single-level design. This fit admirably well
into a rectangular file and was easy to handle using the simple
file-handling capabilities of the SAS System. More complex designs,
involving multiple levels or multiple waves of data gathering,
require much more attention be paid to the design of the files.
Particularly, the flexibility offered by relational database designs
often demands extensive prior planning on the part of the
researcher. Because we did not confront these problems, I shall
leave their description to others.

Scale of Measurement

One major problem was that most multivariate techniques,
including those we used, are not invariant under changes in the
scale of measurement. Both techniques that we used, principal
components and cluster analysis, are known to be strongly influenced
by the scale of measurement.

Unfortunately, the units in which the data were measured varied widely. At one extreme were the percentages used to measure the change variables. For example, population growth, urban primacy, and export dependency on raw materials were all measured as a percentage. These variables have a theoretical range between zero and one hundred. In practice, we usually observed even smaller ranges. For example, the percent of land which is arable ranges from 0.01% (for Iceland) to 66.1% (for Bangladesh) and the percent of manufacturing in GDP ranges from 1.8% (for South Yemen) to 63.8% (for East Germany). At the other extreme are variables like Gross National Product or Trade Turnover whose values have ranges thousands of times greater than the variables measured as percentages. For example, 1975 GNP was $1.52 trillion for the United States, but only $40 million for Tonga--a range of 1 to 38,000. These widely varying ranges were only one problem.

Many times the original scale of the data reflected nothing more than the convenience of the agency that carried out the collection, not any particular brand of economic theory. And scale factors could be a problem. The difference between variables measured in tens of tons per square kilometer (paddy yield) and tons per square kilometer (wheat yield) means only the movement of a decimal point to an analyst. Yet the change in the variance that would result from multiplying paddy yield by ten is significant for the covariance matrices that are the basis of most multivariate analysis techniques.

The statistical literature offered little guidance in this situation. One author simply suggested that these techniques not be employed unless the data were all measured in the same units.[6] We did not find this suggestion particularly helpful since, in most cases, there was a great deal of arbitrariness about the choice of units. For many variables there was no compelling theoretical reason to choose one unit of measurement over another. For example, should gross national product be expressed in dollars or in millions of dollars?[7] Nor is there an obvious choice of units for a variable like paddy yields.[8] Even though the choice of units was not obvious, we could not escape confronting a choice because the computer algorithms used by most multivariate techniques are severely

---

6. A. E. Maxwell, "Factor Analysis: Statistical Aspects," in *International Encyclopedia of the Social Sciences,* ed. David L. Sills, vol. 3, (Chicago: Macmillan & Co., 1968), pp. 280.

7. We chose to use millions of dollars, but mostly for convenience in entering the data into the computer--not as many numbers had to be keypunched.

8. We used tens of tons per square kilometer, again primarily for convenience in data entry.

affected by this choice.

We dealt with this problem by standardizing the variables: all of the multivariate analyses were conducted from the correlation matrix. For our principal components analysis using the correlation matrix had the effect of giving all variables equal weight. Had we used the covariance matrix, the variables would have been weighted in proportion to their variance. Since the variances of our data are extremely heterogeneous and the differences in variance are not theoretically based, use of the covariance matrix would have caused serious distortions in the results.

Skewness

The data for almost every variable were highly skewed, not always in the same direction. For example, gross national product is highly positively skewed, with a skewness coefficient of 8.2 for 180 countries, whereas agricultural production growth, 1970-74, is strongly negatively skewed, with a coefficient of -2.0 for 126 countries.

There are several criteria that can be used to judge the appropriateness of the distribution shown by a variable.[9] Since we did not intend to do any formal hypothesis-testing, we were not concerned that our data be normally distributed or that they have constant variance. We chose to develop empirical transformations of the data which would make each variable as symmetric as possible. This reduction in the extreme skewness present in many of the variables made the means and other statistics--such as variances and covariances--used by the multivariate methods much more accurate as measures of central tendency and spread.

The most common transformation used was the logarithmic transformation. This was used for almost every variable where the ratio of the largest to the smallest value was substantial. Examples include gross national product, population, energy consumption, fertilizer consumption, agricultural yields, and trade turnover, among others. There is an important advantage to using a log scale for these variables. The problem can be best illustrated by using an example which illustrates the problems that cloud interpretation of data which have not been transformed. The basic difficulty is that without the log transformation, the numbers do not have the same meaning at different points along the scale. For

---

9. Cf. M. H. Hoyle, "Transformations - An Introduction and a Bibliography," *International Statistical Review* 41 (1973): 203-223; J. B. Kruskal, "Statistical Analysis: Transformations of Data", in *International Encyclopedia of the Social Sciences,* ed. David L. Sills, vol. 6 (Chicago: Macmillan & Co., 1968), pp. 112-119.

example, a difference of ten million dollars in the gross national product of two countries near the bottom of the scale indicates a significant difference in the productive capacities of their economies--possibly representing a difference of 20 or 25 percent. However, an equal difference in the gross national product of countries near the top of the scale is trivial, well within the margin of error of the reported data. As in this example, when the meaning of differences is proportional to the size of the data on the original scale, the log transformation has the property that the units along all points of the transformed scale have the same meaning. A difference in logged one unit at the bottom represents the same increment in productive capacity as does a difference of one logged unit at the top of the scale. In this sense, the log transformation aids the interpretation of the data because it standardizes the meaning of the units.[10]

Estimation of Missing Data

A fundamental problem for everyone working with cross-national data is the lack of data for many countries. In our case, the problem was extreme. Only four variables (population 1975, arable land per capita, rail route length, and area) were complete for all 185 countries. Furthermore, all of the countries except Brazil contained at least one missing data point.

In many cases, there are excellent reasons for data to be missing. The most common reason is economic. Less developed countries do not have the capabilities for data collection that developed countries have and hence do not collect many kinds of data. Less common reasons include the following: Some countries are closed to the outside world and do not make data public (e.g., North Korea or Albania). In countries involved in wars or civil unrest the data collection systems may break down (e.g., Kampuchea). Finally, some countries simply do not collect certain kinds of data. For example, none of the industrial countries had available data on service on external public debt. Other countries have no need to collect certain statistics. An obvious example: Greenland grows no rice and therefore collects no data on paddy production. There are, however, more subtle examples, such as the lack of data for daily calorie consumption per capita for Ireland or for international mail pieces sent for Uruguay.

---

10. Other transformations used included the arc sine transform for most growth variables and the square root transform for most percentage variables. The variable Raw Materials in Total Exports, Percent, was squared. We do not suggest that these transformations were guided by theory. As the text states, they are *empirical* transforms, chosen to make the distributions of the variables as symmetric as possible.

Missing data are a particular problem for this study for two reasons. First, to avoid the biases of small samples which had not been restricted for theoretical reasons, we attempted to achieve maximum possible coverage. The presence of any missing data reduced the size of the sample for that variable and, potentially, exposed us to the very problems that we were trying to avoid. Second, missing data in cross-national datasets are not randomly distributed. They tend to be concentrated in countries which are small or poor. This creates a problem of bias no matter how the missing data are handled.

Whatever the reason, lack of data poses problems for the researcher. One major problem is statistical. Many data analysis techniques will not accept cases which contain missing data. To avoid this restriction, missing data must be handled differently from non-missing data.

There are several common ways to handle missing data.[11] The most common method, in fact the default method for many computer programs, is to simply delete any case that contains missing data. A major weakness of this solution is that, unless the missing data are randomly distributed (and they were not in our data), it will bias the statistical results. A further problem, particularly acute for us, was that since almost every case had *some* missing data, deletion of these cases would have meant that our analyses would have been based on almost no data. Furthermore, we regarded our countries not as a statistical population of independent, identically distributed observations, but as individual countries which were important in their own right. For these reasons, deleting cases was not a satisfactory general solution. However, in specific cases extremely high levels of missing data compelled us to use it. When countries had extremely high proportions of missing data (e.g., Brunei had over 40% missing data, Luxemburg had over 35%, Western Samoa had over 38%), we deleted them. As a result of this process we eliminated forty-two countries, leaving 143 countries remaining in our multivariate analysis dataset. Countries were never deleted from the univariate analyses, only from the multivariate analyses.

---

11. For discussions of alternative methods for treating missing data, see James W. Frane, "Some Simple Procedures for Handling Missing Data in Multivariate Analysis," *Psychometrika* 41 (1976): 409-415 and E. M. L. Beale and R. J. A. Little, "Missing Values in Multivariate Analysis," *Journal of the Royal Statistical Society,* Series B, 37 (1975): 129-145, pp. 129-145. Much of the succeeding discussion is based on those two sources.

A second method often used in multivariate research is to remove variables with an unacceptably high proportion of missing data from the analysis. We used this method with care. An advantage of this method was that eliminating some variables raised the proportion of non-missing data for most countries and allowed us to retain more countries in the dataset.[12] Furthermore, unlike countries, not all variables were unique. Some variables could be deleted because the dataset contained other variables which were close proxies theoretically and highly correlated empirically. Many of the energy and energy consumption per capita variables fall into this category. These variables could be removed from the multivariate analyses with negligible loss of information. In the end we eliminated 58 variables; leaving 67 variables in our multivariate analysis dataset.

A third method of handling missing data involves the computation of "missing value" correlation or covariance matrices. If the objectives of a multivariate analysis require only the estimation of parameters and not the computation of derived scores, this method may be valuable.[13] However, one of the attractions of the principal components analysis that we employed was that it provided factor scores for individual countries that could be mapped and the patterns of which could be analyzed. In short, the third method was not a viable alternative.

Finally, the researcher can derive estimates for the missing data. We chose to follow this course with the remaining 67 variables and 143 countries. Among the variables, the proportion of missing data ranged from 0% to 27% (for paddy yield). The mean number of countries missing per variable was 8.4%; the median was 4 countries or 2.8%. The countries contained from 0% to 18.9% missing data. The level of missing data was highest for French Guiana with 27 missing variables out of 67. The mean number of missing variables was 8.4%; the median was 2 variables or 3%.

The simplest method for imputing missing data is to replace the missing values by the mean. The use of means, however, can cause severe bias when the wealth and position of a country are such that the mean is an impossible estimate for a missing value. This was true for most of our data, and, on this basis, we rejected this technique.

---

12. The reverse result also occurred as we eliminated countries, with equally favorable results for the variables.

13. By derived scores I mean new variables calculated from the parameter estimates provided by the statistical procedures. Fitted values and residuals from a regression are commonly used derived scores. The derived scores from principal components analysis are called factor scores.

Better results can be obtained by using regression.[14] We
hose to replace the missing values with predicted values obtained
·y using a stepwise regression technique. Each missing value for a
ariable is estimated by regressing that variable on the variables
·hich are significantly associated with it (i.e., meet a standard
.tepwise regression F-to-enter criterion). While this method
nvolves extensive computation and is thus quite costly, it is the
.ost theoretically sound method, "...since it attempts to use the
.aximum amount of information in the available variables in
·stimating the missing variables without overfitting."[15] This method
·as implemented using BMDPAM.[16]

A general comment about bias in techniques that impute
.issing data is in order here. All of the approaches that we know
·equire estimating the covariance matrix of the original data. We
.re aware of no procedure for generating an unbiased estimate of the
:ovariance matrix when the missing data are non-random. But the
)otential bias can be reduced. Thus, the covariance matrix used in
.he regressions that replaced the missing values was estimated from
:he data after transformation and after the dataset had been reduced
.n size by removing variables and countries with large proportions
)f missing data. Because the pattern of missing data was
.on-random, the covariances would still be biased upward.

Analytic Techniques

We chose particular data analytic techniques because they are
.ost appropriate for our goals. Our primary goal in the
.ultivariate analyses was to simplify the process of understanding
.nd interpretation of the information contained in over 100
.ariables. This required techniques which would group variables
into a relatively small number of classes, within which the
:ountries were relatively similar. Statistically this problem is
.nown as the *reduction of dimensionality*.[17]

There are many different techniques available. Which one is
.ost appropriate depends on the type of data and the goals of the
.esearch. In this case, our work was descriptive and exploratory.
.e were not working deductively from a well-defined, existing
:heory. Instead our interest focussed on the systematic

---

14. See Frane, "Some Simple Procedures for Handling Missing Data in
Multivariate Analysis" for a discussion of several regression techniques.

15. Ibid., p. 411.

16. W. J. Dixon, *BMDP Biomedical Computer Programs P-Series,* (Berkeley:
University of California Press, 1975).

17. Gnanadesikan, *Methods for Statistical Data Analysis of Multivariate
Observations.*

relationships which would be revealed by the data themselves. Thus the multivariate techniques that were most useful were those in which variables are grouped in ways determined from the data themselves.

We were interested not only in a convenient empirical reduction in the amount of data, but, more important, we hoped to find underlying types or classes of countries. This would facilitate further analysis in that it would suggest underlying similarities in the current status and trajectory of growth that countries are following. This is of scholarly interest because of the light it casts on the match between theories of economic growth and ways in which growth is actually occurring in the world today. It is of interest to policy makers because it may help clarify both the strategy that a country will need to follow to assist its economic growth and the strategy that nations and international agencies should follow in providing development assistance.

The multivariate analyses were conducted using two primary techniques: Principal components analysis and cluster analysis. These two techniques are discussed below, followed by a comparison of the similarities and differences of the techniques. To supplement these techniques, we constructed a set of "World Standard Indices," and we made extensive use of maps. A discussion of these methods follows the comparison of principal components and clustering.

Principal Components Analysis

Principal components analysis, a form of factor analysis, is a standard tool in the analysis of patterns of development.[18] As with all forms of factor analysis, principal components "assumes that the observed variables are linear combinations of some underlying (hypothetical or unobservable) factors."[19] The components are derived so that the first principal component accounts for the maximum variance; the second component accounts for the maximum amount of the remaining variance subject to the constraint that it must be uncorrelated with the first component, etc. The correlation of each observed variable with each component (called the factor loading) represents the extent to which that variable is associated with that component. If a component is interpreted as representing

---

18. Cf. J. H. F. Schilderinck, *Factor Analysis Applied to Developed and Developing Countries,* (Netherlands: Rotterdam University Press, 1969) or R. J. Rummel, *The Dimensions of Nations,* (Beverly Hills, CA: Sage Publications, 1972).

19. Jae-On Kim and Charles W. Mueller, *Introduction to Factor Analysis: What It Is and How to Do It,* Sage University Paper series on Quantitative Applications in the Social Sciences, 07-013 (Beverly Hills: Sage Publications, 1977), p. 8.

he underlying source variable, then the factor loading represents he degree to which the variable is associated with the source variable. A high factor loading (correlation) means a strong association; a low loading suggests that the variable is weakly associated with the source variable.[20]

Principal components is particularly appropriate for this research because of the way it chooses the components that it extracts from the data. First, unlike the common factor model of factor analysis, it requires no assumptions about the underlying distribution of the observed variables.[21] Second, we could obtain factor score coefficients from the components to be mapped and studied like other variables in the dataset. There is no statistical procedure for estimating scores for subjects from the common factor model. This problem, called "factor indeterminacy," arises because the common factor model is based on more unobserved parameters than observed data points.[22] Third, principal components does not require imposing a causal model on the data in order to extract the components. Instead, it chooses components based on the amount of variance that they explain.[23] It requires fewer prior assumptions about what those patterns will be. For exploratory work like ours, this seems more fruitful since it allows the researcher to be more open to the patterns revealed by the data.

Several comments are in order concerning the principal components analyses that we performed. Following the extraction of the principal components we rotated the axes using a varimax orthogonal rotation.[24] Varimax rotation was chosen because the evidence seems to indicate that it gives clearer separation of the components than do alternative orthogonal rotation techniques. While most of our analyses sought a simple subset of the variables, we did an analysis that attempted to find groupings of countries. In general we relied on the scree test and the criterion of

---

20. Readers who would like a deeper understanding of factor analysis in general and principal components analysis in particular are urged to consult Kim and Mueller, *Introduction to Factor Analysis* or Harry H. Harmon, *Modern Factor Analysis,* 3rd Ed. (Chicago: University of Chicago Press, 1976). For a non-technical history of factor analysis, focusing mainly on uses in psychology, see S. J. Gould, *The Mismeasurement of Man,* (New York: W. W. Norton & Co., 1981).

21. A. E. Maxwell, "Factor Analysis: Statistical Aspects," in *International Encyclopedia of the Social Sciences,* ed. David L. Sills, vol. 3, (Chicago: Macmillan & Co., 1968), pp. 276-281.

22. For a general review see W. W. Rozeboom, "The Determinacy of Common Factors in Large Item Domains," *Psychometrika* 47 (1982) 281-295. Note that some computer programs provide "regression" estimates, but these are not estimates in the usual statistical sense.

23. Kim and Mueller, *Introduction to Factor Analysis,* pp. 19-20.

24. Henry F. Kaiser, "The Varimax Criterion for Analytic Rotation in Factor Analysis," *Psychometrika* 23 (1958): 187-200.

interpretability to determine the number of factors.[25] The computer
program which performed the actual analyses was BMDP4M.[26]

Hierarchical Cluster Analysis

To check the results of the principal components analysis, w
used a hierarchical cluster analysis of the variables. A systemati
discussion comparing cluster analysis with principal components
appears in the following section.

The purpose of cluster analysis is to place variables into
groups (called clusters) suggested by the data, not defined a
priori. Clusters have the property that variables in a given
cluster tend to be similar to each other in some sense, and
variables in different clusters tend to be dissimilar.

There are a wide variety of clustering methods available.[27]
We chose a hierarchical clustering technique for two reasons.
First, it offers an essentially unlimited choice of groups. The
entire clustering process is openly displayed. We could easily
examine alternative groupings of the data as well as sub-clusters o
individual clusters. Given the empirical, exploratory nature of
this research, this was an important advantage. Second,
hierarchical clustering requires that very little be assumed prior
to the computation of the clusters. Decisions about how many
clusters or how many levels of clusters were postponed until we
actually looked at the output from the computer program.[28] The only
prior decision to be made is what to use as the measure of distance
between two clusters of any size. We chose the maximum method (als
called the complete linkage method) for calculating distances
because it seems to yield smaller, more compact clusters.[29]

The number of clusters that form the "best" solution is very
difficult to determine.[30] If anything, this problem is more
difficult than the number of components problem in principal

---

25. For the scree test see Raymond B. Cattell, "The Scree Test for the
Number of Factors," *Multivariate Behavioral Research* 1 (1966): 245-276 and H. L.
Horn and R. Engstrom, "Cattell's Scree Test in Relation to Bartlett's Chi-squar
Test and Other Observations on the Number of Factors Problem," *Multivariate
Behavioral Research* 14 (1979): 283-300. For information on the criterion of
interpretability see C. W. Harris, "On Factors and Factor Scores," *Psychometrika*
32 (1967): 363-379.

26. Dixon, *BMDP Biomedical Computer Programs P-Series.*

27. See B. S. Everitt, *Cluster Analysis,* 2nd ed. (London: Heineman
Educational Books Ltd., 1980), for a general introduction to cluster analysis.

28. David Wallace, "Clustering", in *International Encyclopedia of the
Social Sciences,* ed. David L. Sills, vol. 2 (Chicago: Macmillan & Co. 1968), p
519-524.

29. Gnanadesikan, *Methods for Statistical Data Analysis of Multivariate
Observations,* p. 110.

30. Everitt, *Cluster Analysis.*

omponents analysis. Since clustering algorithms attempt to
aximize the distance between clusters, ordinary significance tests
re not valid for testing differences between clusters. Although
everal heuristic methods have been suggested, we know of no
heoretically sound methods for judging the number of clusters.
his is particularly true if a researcher does not feel justified in
aking any distributional assumptions. The dendrogram (also called
 dendogram, a linkage, tree or a phenogram) and the human eye are
he most powerful tools currently available. The actual cluster
nalysis was BMDP1M, cluster analysis of variables.[31]

The categories resulting from the cluster analysis did not
enerally differ from the components from the principal components
nalysis. This was a comforting validation of the results of the
rincipal components. Because the results did not differ, the
luster analysis is not presented in the text.

Comparison of Principal Components and Cluster Analysis

A brief comparative summary of the two techniques will help
o clarify the strengths and weaknesses of each.

Similarities:

1. Both methods are attempts to discover structure in
    an unstructured rectangular dataset.

2. Neither method requires any distribution assumptions.

3. The results of both methods depend strongly on
    which variables are included in the input.

4. The results of both methods are influenced by linear
    changes in the scale of measurement--e.g., whether
    gross national product is measured in dollars or in
    millions of dollars. In our work, scale-of-
    measurement effects were removed by doing all
    analyses on the correlation matrices.

5. Both methods have no clear-cut way of determining the
    number of components/clusters.

Differences:

1. In conventional statistical terms the major
    difference between the two methods is as follows:
    Principal components is an attempt to find a few
    statistically uncorrelated, "independent" variables
    such that regression on those variables fits the

---

31 . Dixon, *BMDP Biomedical Computer Programs P-Series.*

observed data. By contrast, cluster analysis is an
attempt to classify the variables into a few
categories such that these categories capture the
major differences between the variables. In this
sense, cluster analysis is similar to an analysis
of variance classification. [32]

2. Cluster analysis may be carried out on a finer
scale than principal components. By using the
dendrogram to divide classes into subclasses an
entire taxonomy may be obtained.

3. Principal components analysis has the capability to
compute derived scores for individual countries.
These scores can be used as input into other
statistical procedures and can be plotted, regressed
or ANOVAed.

In general, the principal components analyses were our major
tool. In order to do the mapping, the scores for individual
countries given by the factor score coefficients were indispensable

The taxonomy feature of the cluster analysis proved very
helpful in one instance. In our major analyses, the first principal
component seemed to be interpretable as a very general economic
development factor. The generality of this component was
disconcerting since our goal was to uncover patterns of theoretical
and policy significance. An examination of the corresponding
cluster in the cluster analysis suggested that there were strong
sub-clusters. This suggested significant variation among the
variables loading heavily on that component. Since the most
influential decision that a researcher makes concerns the variables
to be used as input and since the first component represents nothing
more than the axis along which there was the greatest variation in
the data, we decided to analyze the subset of the variables which
were strongly correlated with that component. The cluster analysis
suggested that the subset contained several distinguishable
components that would be less general and, hence, more instructive.
The point here is that, by removing the variables that were not a
part of that component, we removed much of the noise that the
computer program had to deal with in the larger analysis. The

32 . Wallace, "Clustering," p. 523.

:sults of this exercise were very informative, and they can be seen
1 chapter 2, figure 1.

World Standard Indices

Principal components and cluster analysis combine variables
)r countries) based on similarity, but this is not the only
1eoretically meaningful way to combine variables. Highly
lssimilar variables can be combined into an index that is
1eoretically valuable and useful as an indicator for policymakers.
1ere are many ways to combine variables into an index. One
iccessful approach was used by the Overseas Development Council to
:eate an index called the Quality of Life Index.33 We followed a
imilar approach to develop a series of indices, called World
:andard Indices, to measure other variables of interest. These
idices were discussed in chapter 3 as one of the "Alternatives to
)nventional Multivariate Analysis."

The following procedure was used to create each index. Begin
v identifying the countries with the highest factor scores on the
:onomic Development per capita component of the factor analysis
?actor 1, see table 2). Countries with high scores on Economic
evelopment per capita but with populations of under ten million
2.g., Iceland or the Bahamas) or with large oil export wealth
2.g., Kuwait or Venezuela) were eliminated. There remained
2venteen nations: Australia, Austria, Belgium, Canada, Denmark,
inland, France, Israel, Japan, Netherlands, New Zealand, Norway,
veden, Switzerland, United Kingdom, United States of America, and
est Germany. As you can, see this category includes basically the
ntire Western industrialized world.

Next, the variables that would compose an index were
elected. This is the most crucial step. The selection criteria
re discussed in chapter III. Once the variables were selected, we
alculated the mean of the values of the seventeen nations named
bove for each variable in the index and set that average to 100.
his formed the "world standard" in the sense that it represented
enerally the highest attainable value for that variable in the
orld today. This transformation was applied to all of the
ountries for each variable in the index, resulting in a new set of
ransformed variables. To create the index itself, we calculated
he mean of the transformed variables for each country. Countries
ith missing data for half or more of the variables in the index
ere omitted. For other countries with less than half of the

---

33 . Overseas Development Council, op. cit.

variables missing, the mean was based on only those variables for which data were present.

The effects and characteristics of the calculation of a wor standard and the scale that it creates are important. The effect this transformation is to create a standardized scale independent the scale on which the variable was measured. Thus the transformation facilitates direct comparisons of countries across many different variables. In this sense, this scaling process is similar to the six-category scale used to draw the maps. It is, o course, different in many other respects. The scale is an interva scale with both zero and 100 being strictly defined. For most variables, the range of scores was from near zero to slightly above 100. Since the world standard was 100 for each of the variables, the world standard of the average (i.e., the index) will also be 100. Note that missing data will bias the scale values in the following way. Since the variables making up the scale are often diverse, it is possible for a country to rank high on some and low on others. If the variables on one end of the ranking, say all the low rankings, are missing, then the country will be ranked disproportionately high since the high ranking variables will stil be included in the index. Missing data had virtually no effect on the world standard since the countries from which it was calculate rarely had any missing values.

The value of this technique lies in the greater control that it gives the researcher over the variables included in the index. An example of a situation where this enabled us to overcome a weakness of both principal components and cluster analysis is the Growth Potential Index. Principal components and cluster analysis base their results, by necessity, on historical data, but the growt potential of a country is a question dealing with the future. The data are not yet available. While there is certainly some degree c continuity between past performance and future growth, there is als some discontinuity. The Growth Potential Index gave us a way of looking at the future without being as closely tied to the past as other techniques were.

Maps

Although they are not usually thought of as a multivariate technique, we have used maps to illustrate and further refine the results of the principal components analyses. Maps are underutilized in social science research. They are potentially so useful that they deserve a brief commentary.[34]

Maps, and other graphical displays, have the unique characteristic that they allow readers to see individual countries and overall patterns. Much of the discussion of the maps reflected this twin search for patterns as well as individual countries which deviated from the patterns in unexpected ways. This characteristic is what made maps such an excellent data analytic tool.

Several other characteristics of the maps makes them valuable tools. If well designed, they can present large datasets in a small space without becoming confusing or incoherent. They encourage your eye to compare different countries. Maps take advantage of the fact that our eye-brain system is the most sophisticated information processor ever developed. Through cartographic displays we put this system to use to obtain deep insight into the structure of global wealth and poverty.[35]

Summary

The methodological decisions that we made were governed by the conviction that choice of methods is important. Guiding our choice of techniques was the combination of (1) the goals of the research and (2) our understanding of nature of the phenomenon under study. The goals of the research restricted our choice of variables to those influencing economic growth and development. For those variables, we sought maximum coverage. To avoid small sample bias and non-theoretically grounded restrictions on sample size, we gathered data for as many nations as possible.

Once collected, the data required a great deal of preparation before the multivariate analyses could begin. To make the many highly skewed variables symmetric, we developed empirical transformations. The units in which a variable is measured are not theoretically interesting, even though they effect statistical methods. This influence was eliminated by conducting all analyses from the correlation matrix. Finally, to maintain maximum coverage despite the usual number of missing values, we used a regression technique to estimate replacements for the missing data points.

Our research was exploratory and we chose four complementary analytic techniques which were open to the patterns revealed by the data. The key point is not the number of techniques, but the

---

34 . For an excellent discussion of data maps and graphics in general see Edward Tufte, *The Visual Display of Quantitative Information* (Connecticut: Graphics Press, 1983), especially pages 16-27. Much of the following discussion follows Tufte.

35 . Paraphrased from John M. Chambers, William S. Cleveland, Beat Kleiner, and Paul A. Tukey, *Graphical Methods for Data Analysis* (Boston: Duxbury 1983), p. 1.

complementary nature of the different methods. Each was used to offset the weaknesses of the others. The primary technique, principal components analysis, produced the components that had the key role in the reduction in dimensionality of the dataset. It also generated factor score coefficients used to study individual countries. Hierarchical cluster analysis supplemented our understanding of the meaning of the components. The World Standard Indices enabled us to combine variables and generate maps based on other criteria than similarity. Finally, the maps and figures gave us the ability to move from the analysis of variables to examine individual countries and groups of countries.

APPENDIX A
LIST OF COUNTRIES BY REALMS

The table below lists all 185 countries for which data were collected. The well-informed reader will know that this is more countries than there are in the world. The reason for this is that we collected data for a number of administrative entities which are not countries. French Polynesia, Greenland and Puerto Rico are some examples. The 143 countries included in the multivariate analyses are marked with an asterisk.

Realm: Australasia
1. Australia*
2. Fiji*
3. French Polynesia
4. Gilbert Islands
5. Guam
6. New Caledonia
7. New Hebrides
8. New Zealand*
9. Pacific Islands, U.S. Trust
10. Papua New Guinea*
11. Samoa, American
12. Samoa, Western
13. Solomon Islands
14. Tonga

Realm: Anglo-America
1. Bermuda
2. Canada
3. Greenland
4. U.S.A.*

Realm: Latin America
1. Antigua
2. Argentina*
3. Bahamas*
4. Barbados
5. Belize*
6. Bolivia*

7. Brazil*
8. Chile*
9. Colombia*
10. Costa Rica*
11. Cuba*
12. Dominica
13. Dominican Republic*
14. Ecuador*
15. El Salvador*
16. French Guiana*
17. Grenada
18. Guadeloupe*
19. Guyana*
20. Haiti*
21. Honduras*
22. Jamaica*
23. Martinique*
24. Mexico*
25. Netherlands Antilles
26. Nicaragua*
27. Panama*
28. Panama Canal Zone
29. Paraguay*
30. Peru*
31. Puerto Rico
32. St. Kitts-Nevis
33. St. Lucia
34. St. Vincent
35. Surinam*
36. Trinidad*
37. Uruguay*
38. Venezuela*
39. Virgin Islands

Realm: Western Europe

1. Austria*
2. Belgium*
3. Denmark*
4. Finland*
5. France*
6. Germany, West*
7. Greece*

8. Iceland*
9. Ireland*
10. Italy*
11. Liechtenstein
12. Luxembourg
13. Malta*
14. Monaco
15. Netherlands*
16. Norway*
17. Portugal*
18. Spain*
19. Sweden*
20. Switzerland*
21. United Kingdom*

Realm: Eastern Europe

1. Albania*
2. Bulgaria*
3. Czechoslovakia*
4. Germany, East*
5. Hungary*
6. Poland*
7. Rumania*
8. U.S.S.R.*
9. Yugoslavia*

Realm: Near East

1. Algeria*
2. Bahrain*
3. Cyprus*
4. Egypt*
5. Iran*
6. Iraq*
7. Israel*
8. Jordan*
9. Kuwait*
10. Lebanon*
11. Libya*
12. Morocco*
13. Oman
14. Qatar
15. Sahara, Western
16. Saudi Arabia*

17. Syria*
18. Tunisia*
19. Turkey*
20. United Arab Emirates
21. Yemen, North*
22. Yemen, South*

Realm: Sub-Saharan Africa

1. Angola*
2. Benin*
3. Botswana*
4. Burundi*
5. Cameroon*
6. Cape Verde
7. Central African Empire*
8. Chad*
9. Comoro Islands
10. Congo*
11. Djibouti
12. Equatorial Guinea*
13. Ethiopia*
14. Gabon*
15. Gambia*
16. Ghana*
17. Guinea*
18. Guinea Bissau*
19. Ivory Coast*
20. Kenya*
21. Lesotho*
22. Liberia*
23. Madagascar*
24. Malawi*
25. Mali*
26. Mauritania*
27. Mauritius*
28. Mozambique*
29. Namibia
30. Niger*
31. Nigeria*
32. Reunion*
33. Rhodesia*
34. Rwanda*

35. Sao Tome
36. Senegal*
37. Seychelles
38. Sierra Leone*
39. Somalia*
40. South Africa*
41. Sudan*
42. Swaziland*
43. Tanzania*
44. Togo*
45. Uganda*
46. Upper Volta*
47. Zaire*
48. Zambia*

Realm: South Asia

1. Afghanistan*
2. Bangladesh*
3. Bhutan
4. India*
5. Maldive Islands
6. Nepal*
7. Pakistan*
8. Sri Lanka*

Realm: East Asia

1. China, Peoples Republic of*
2. China, Republic of (Taiwan)*
3. Hong Kong*
4. Japan*
5. Korea, Democratic Peoples Republic of*
6. Korea, Republic of*
7. Macau
8. Mongolia*

Realm: Southeast Asia

1. Brunei
2. Burma*
3. Indonesia*
4. Kampuchea, Democratic (Cambodia)*
5. Laos*
6. Malaysia*
7. Philippines*

8. Singapore*
9. Thailand*
10. Timor, East
11. Vietnam

APPENDIX B
LIST OF VARIABLES

The table below lists all 163 variables for which we collected data. Listed first are the 125 variables relating to general economic development. The second section contains the 40 variables bearing on international development assistance. The table includes variables which were keypunched, e.g., Gross National Product or Population, as well as variables constructed from the keypunched data, e.g., Gross National Product per capita. The 66 variables used in the multivariate analyses have been marked with an asterisk.

1. Air freight transported, domestic, 1975, 100,000 ton-kilometers
2. Air kilometers flown, domestic, 1975, 100,000 kilometers flown
3. Air kilometers per capita, domestic, 1975, kilometers flown per 100 people
4. Air passengers, domestic, 1974, millions of passengers per kilometer
5. Agricultural population, percent of economically active population*
6. Agricultural population 1965
7. Agricultural population 1975
8. Agricultural population, 1975, percent of total population*
9. Agricultural population growth, 1965-75, mean yearly percent change*
10. Agricultural population growth, 1960-70, mean yearly percent change
11. Agricultural production growth, 1970-74, mean yearly percent change
12. Agriculture as percent of Gross Domestic Product, 1973, current factor cost*
13. Arable land, percent of total land area
14. Arable land, 1975, square kilometers
15. Arable land per capita, 1975, square kilometers per person*
16. Arable land per agricultural population, 1975, square kilometers per thousand people*
17. Area, total, 1975, square kilometers*

18. Calories per capita per day*
19. Consumer Price Index Change, 1970-75, all items (1970=100)
20. Consumer Price Index Change, 1970-75, food, (1970=100)
21. Debt, service on external public debt, 1970s, percent of exports
22. Energy Consumption: commercial consumption, 1965, 10,000 tons coal equivalent
23. Energy Consumption: commercial consumption, 1975, 10,000 tons coal equivalents*
24. Energy Consumption: commercial consumption as percent of gross consumption, 1975*
25. Energy Consumption: electricity generation, 1975, millions of kilowatt hours
26. Energy Consumption: fuel wood consumption, 1975, thousands of metric tons coal equivalents*
27. Energy Consumption: gross consumption, 1975, thousands of metric tons coal equivalents*
28. Energy Consumption: hydroelectric generation, 1975, millions of kilowatt hours
29. Energy Consumption growth: Commercial consumption 1965-75, mean percent per year
30. Energy Consumption per capita: hydroelectric generation, 1975, kilowatt-hours per person*
31. Energy Consumption per capita: electricity generation, 1975, kilowatt-hours per person*
32. Energy Consumption per capita: Commercial consumption, 1965, tons of coal equivalents per 100 people*
33. Energy Consumption per capita: Commercial consumption, 1975, tons of coal equivalents per 100 people*
34. Energy Consumption per capita: gross consumption, 1975, tons coal equivalents per 10 persons*
35. Energy Consumption per capita growth: Commercial consumption, 1965-75, mean percent per 100 people per year*
36. Energy Density, total energy potential per square kilometer, 1975, tons of coal equivalents per square kilometer
37. Energy Potential: 200 year biomass potential, millions of tons coal equivalents
38. Energy Potential: non-renewable potential, 1970s, millions of tons coal equivalents
39. Energy Potential: renewable energy potential, 1970s, millions of tons coal equivalents

40. Energy Potential: total energy potential, 1975, millions of tons of coal equivalents
41. Energy Potential per capita: 200 year biomass potential, tons of coal equivalents per person*
42. Energy Potential per capita: non-renewable energy potential, 1970s, tons of coal equivalents per capita
43. Energy Potential per capita: renewable energy potential, 1970s, tons of coal equivalents per person
44. Energy Potential per capita: total energy potential, 1975, tons of coal equivalents per capita
45. Energy Reserves: natural gas reserves, 1975, billions of cubic meters
46. Exports, 1975, percent of Gross National Product*
47. Export Concentration, 1972, Hirschman normalized index*
48. Exports of raw materials, 1973, percent of total exports*
49. Exports to industrialized countries, 1975, percent of total exports*
50. Fertilizer Consumption, 1975, kilograms per square kilometer of arable land*
51. Fertilizer Consumption per capita, 1975, kilograms per 10 agricultural population*
52. Gross Domestic Product, 1975, 100,000 US dollars
53. Gross Domestic Product per capita, 1975, US dollars per person
54. Gross National Product, 1965, 100,000 US dollars
55. Gross National Product per capita, 1975, 100,000 US dollars*
56. Gross National Product per capita, 1965, US dollars
57. Gross National Product per capita, 1975, US dollars*
58. Gross National Product per capita growth, 1960-75, mean percent per year*
59. Gross National Product per capita growth, 1970-75, mean percent per year*
60. Infant Mortality, 1970, deaths per thousand live births*
61. Investment: Gross Domestic Investment as percent of Gross Domestic Product, 1973*
62. Investment: Growth in Gross Domestic Investment, 1965-73, mean percent per year
63. Imports, 1975, millions of US dollars
64. Imports of industrial countries, 1975, millions of US dollars
65. Life Expectancy at birth, 1970s*
66. Literacy Rate, adult, 1970, percent*
67. Mail Flow, international, 1975, pieces sent*

68. Mail Flow per capita: international, 1975, pieces sent per thousand people*
69. Manufacturing percent of Gross Domestic Product, 1973, current factor cost*
70. Newspaper Circulation, 1970s, hundreds of copies daily
71. Newspaper Circulation per capita: 1970s, daily copies per thousand people*
72. Paddy Yield, 1975, tens of metric tons per square kilometer*
73. Passengers, air and rail, 1974, millions of passenger kilometers
74. Passengers per capita, air and rail, 1975, passenger kilometers per person*
75. Physicians & Dentists, 1970s
76. Physicians & Dentists, per 100,000 people, 1970s*
77. Population, 1965
78. Population, mid-1975*
79. Population Density, 1965, people per square kilometer
80. Population Density, 1975, people per square kilometer*
81. Population Growth, 1965-75, mean percent per year*
82. Radio Ownership per capita, 1973-74, radios per thousand people*
83. Rail freight, 1975, millions of ton-kilometers
84. Rail freight per capita, 1975, ton-kilometers per million persons
85. Rail freight per route length, 1975, thousands of ton-kilometers per kilometer of rail route length*
86. Rail passengers, 1974, millions of passenger kilometers
87. Rail route density, 1975, kilometers per thousand of square kilometers*
88. Rail route length, 1975, kilometers
89. Rail route length per capita, 1975, kilometers per 100,000 people*
90. Road freight, 1975, millions of ton-kilometers
91. Road network length, 1975, kilometers
92. Road network length per capita, 1975, kilometers per 100,000 people*
93. Road network density, 1975, kilometers per thousand of square kilometers*
94. School enrollment, primary, gross enrollment ratio, 1960, percent
95. School enrollment, primary, gross enrollment ratio, 1970, percent*

96. School enrollment change, primary, 1960-70, gross percent
enrollment growth*
97. School enrollment, secondary, gross ratio, 1960, percent
98. School enrollment, secondary, gross ratio, 1970, percent*
99. School enrollment, third level, gross ratio, 1974, percent*
100. Steel consumption, 1975, thousands of metric tons
101. Steel consumption per capita, 1975, tons per thousand
people*
102. Tractors, 1975
103. Tractors per capita, 1975, tractors per million persons
agricultural population*
104. Tractors per square kilometer of arable land, 1975*
105. Tractors per paddy yield, 1975, tractors per kilogram
106. Tractors per wheat yield, 1975, tractors per kilogram
107. Trade turnover, 1965, millions of US dollars
108. Trade turnover, 1975, millions of US dollars*
109. Trade turnover growth, 1965-75, mean percent per year*
110. Trade turnover per capita, 1965, US dollars per person
111. Trade turnover per capita, 1975, US dollars per person*
112. Urban population, 1975, people in cities over 100,000
people???
113. Urban population, 1975, percent in cities over 100,000
people*
114. Urban population 1964, percent of total population, census
definition
115. Urban population, 1970s, percent of total population, census
definition*
116. Urban population, 1970s, primacy, percent*
117. Urban population change, 1964-70, mean yearly percent change
118. Vehicles, 1974, total motor vehicles
119. Vehicles, 1975, commercial motor vehicles
120. Vehicles per capita, 1974, motor vehicles per thousand
people*
121. Vehicles per capita, 1974, commercial motor vehicles per
100,000 people*
122. Vehicles per road length, 1974, motorized vehicles per 100
kilometers*
123. Water goods transported, inland, 1975, millions of
ton-kilometers
124. Wheat yield, 1975, metric tons per square kilometer*
125. Youthfulness, 1970s, percent of population age 14 or less*

International Development Assistance

1. Assistance, centrally planned economies, bilateral 1969-75, millions of US dollars
2. Assistance, International Bank for Reconstruction and Development (World Bank), 1946-76, 100,000 US dollars
3. Assistance, total international organizations, 1946-76, 100,000 US dollars
4. Assistance, DAC and multilateral net concessional, 1969-75, millions of US dollars
5. Assistance, DAC and multilateral total net, 1969-75, millions of US dollars
6. Assistance, multilateral agricultural development, 1966-1976, millions of US dollars
7. Assistance, multilateral commodities development, 1966-1976, millions of US dollars
8. Assistance, multilateral infrastructure development, 1966-1976, millions of US dollars
9. Assistance, multilateral industrial develoment, 1966-1976, millions of US dollars
10. Assistance, multilateral social development, 1966-1976, millions of US dollars
11. Assistance, multilateral total, 1966-1976, millions of US dollars
12. Assistance, United Nations, 1946-1976, 100,000 US dollars
13. Assistance U.S.A. economic, 1946-1976, 100,000 US dollars
14. Assistance U.S.A. economic, 1962-1976, 100,000 US dollars
15. Assistance, U.S.A. AID & PL480 agricultural development, 1966-1976, millions of US dollars
16. Assistance, U.S.A. AID & PL480 commodities, 1966-1976, millions of US dollars
17. Assistance, U.S.A. AID & PL480 infrastructure development, 1966-76, millions of US dollars
18. Assistance, U.S.A. AID & PL480 industrial development, 1966-1976, millions of US dollars
19. Assistance, U.S.A. AID & PL480 social development, 1966-1976, millions of US dollars
20. Assistance, U.S.A. AID & PL480 total, 1966-1976, millions of US dollars
21. Assistance per capita: centrally planned economies, bilateral, 1969-1975
22. Assistance per capita: DAC multilateral net concessional, 1969-1975, US dollars per person

23. Assistance per capita: DAC multilateral total net, 1969-1975, US dollars per person
24. Assistance per capita: total international organizations, 1946-76, 1965 population
25. Assistance per capita: multilateral agricultural development, 1966-1976, US dollars per person
26. Assistance per capita: multilateral commodities development, 1966-1976, US dollars per person
27. Assistance per capita: multilateral infrastructure development, 1966-1976 US dollars per person
28. Assistance per capita: multilateral industrial development, 1966-1976, US dollars per person
29. Assistance per capita: multilateral social development, 1966-1976, US dollars per person
30. Assistance per capita: multilateral total, 1966-1976, US dollars per person
31. Assistance per capita, U.S.A. economic, 1946-1976, US dollars per person (1965 population)
32. Assistance per capita: U.S.A. economic, 1962-1976, US dollars per person (1965 population)
33. Assistance per capita: U.S.A. AID & PL480 agricultural development, 1966-1976
34. Assistance per capita: U.S.A. AID & PL480 commodities, 1966-1976
35. Assistance per capita: U.S.A. AID & PL480 infrastructure development, 1966-1976
36. Assistance per capita: U.S.A. AID & PL480 industrial development, 1966-1976
37. Assistance per capita: U.S.A. AID & PL480 social development, 1966-1976
38. Assistance per capita: U.S.A. AID & PL480 total, 1966-1976
39. Assistance per capita: U.S.A. economic, 1946-1976, 1965 population
40. Assistance per capita: U.S.A. economic, 1962-1976, 1965 population

THE UNIVERSITY OF CHICAGO
DEPARTMENT OF GEOGRAPHY
RESEARCH PAPERS (Lithographed, 6×9 inches)

LIST OF TITLES IN PRINT

133. SCHWIND, PAUL J. *Migration and Regional Development in the United States.* 1971. 170 p.

134. PYLE, GERALD F. *Heart Disease, Cancer and Stroke in Chicago: A Geographical Analysis w Facilities, Plans for 1980.* 1971. 292 p.

135. JOHNSON, JAMES F. *Renovated Waste Water: An Alternative Source of Municipal Water Sup, in the United States.* 1971. 155 p.

136. BUTZER, KARL W. *Recent History of an Ethiopian Delta: The Omo River and the Level of L(Rudolf.* 1971. 184 p.

139. MCMANIS, DOUGLAS R. *European Impressions of the New England Coast, 1497–1620.* 1972. 14

140. COHEN, YEHOSHUA S. *Diffusion of an Innovation in an Urban System: The Spread of Plan; Regional Shopping Centers in the United States, 1949–1968,* 1972. 136 p.

141. MITCHELL, NORA. *The Indian Hill-Station: Kodaikanal.* 1972. 199 p.

142. PLATT, RUTHERFORD H. *The Open Space Decision Process: Spatial Allocation of Costs a Benefits.* 1972. 189 p.

143. GOLANT, STEPHEN M. *The Residential Location and Spatial Behavior of the Elderly: A Canadt Example.* 1972. 226 p.

144. PANNELL, CLIFTON W. *T'ai-chung, T'ai-wan: Structure and Function.* 1973. 200 p.

145. LANKFORD, PHILIP M. *Regional Incomes in the United States, 1929–1967: Level, Distributic Stability, and Growth.* 1972. 137 p.

146. FREEMAN, DONALD B. *International Trade, Migration, and Capital Flows: A Quantitative An ysis of Spatial Economic Interaction.* 1973. 201 p.

147. MYERS, SARAH K. *Language Shift Among Migrants to Lima, Peru.* 1973. 203 p.

148. JOHNSON, DOUGLAS L. *Jabal al-Akhdar, Cyrenaica: An Historical Geography of Settlement a Livelihood.* 1973. 240 p.

149. YEUNG, YUE-MAN. *National Development Policy and Urban Transformation in Singapore: Study of Public Housing and the Marketing System.* 1973. 204 p.

150. HALL, FRED L. *Location Criteria for High Schools: Student Transportation and Racial In gration.* 1973. 156 p.

151. ROSENBERG, TERRY J. *Residence, Employment, and Mobility of Puerto Ricans in New York Ci* 1974. 230 p.

152. MIKESELL, MARVIN W., editor. *Geographers Abroad: Essays on the Problems and Prospec of Research in Foreign Areas.* 1973. 296 p.

153. OSBORN, JAMES F. *Area, Development Policy, and the Middle City in Malaysia.* 1974. 291 p.

154. WACHT, WALTER F. *The Domestic Air Transportation Network of the United States.* 1974. 98

155. BERRY, BRIAN J. L., *et al. Land Use, Urban Form and Environmental Quality.* 1974. 440 p.

156. MITCHELL, JAMES K. *Community Response to Coastal Erosion: Individual and Collective A justments to Hazard on the Atlantic Shore.* 1974. 209 p.

157. COOK, GILLIAN P. *Spatial Dynamics of Business Growth in the Witwatersrand.* 1975. 144 p.

159. PYLE, GERALD F. *et al. The Spatial Dynamics of Crime.* 1974. 221 p.

160. MEYER, JUDITH W. *Diffusion of an American Montessori Education.* 1975. 97 p.

161. SCHMID, JAMES A. *Urban Vegetation: A Review and Chicago Case Study.* 1975. 266 p.

162. LAMB, RICHARD F. *Metropolitan Impacts on Rural America.* 1975. 196 p.

163. FEDOR, THOMAS STANLEY. *Patterns of Urban Growth in the Russian Empire during the Nin teenth Century.* 1975. 245 p.

164. HARRIS, CHAUNCY D. *Guide to Geographical Bibliographies and Reference Works in Russic or on the Soviet Union.* 1975. 478 p.

165. JONES, DONALD W. *Migration and Urban Unemployment in Dualistic Economic Developmen* 1975. 174 p.

166. BEDNARZ, ROBERT S. *The Effect of Air Pollution on Property Value in Chicago.* 1975. 111 p.

167. HANNEMANN, MANFRED. *The Diffusion of the Reformation in Southwestern Germany, 151& 1534.* 1975. 248 p.

168. SUBLETT, MICHAEL D. *Farmers on the Road. Interfarm Migration and the Farming of No. contiguous Lands in Three Midwestern Townships. 1939–1969.* 1975. 228 pp.

169. STETZER, DONALD FOSTER. *Special Districts in Cook County: Toward a Geography of Loc Government.* 1975. 189 pp.

170. EARLE, CARVILLE V. *The Evolution of a Tidewater Settlement System: All Hallow's Paris. Maryland, 1650–1783.* 1975. 249 pp.

171. SPODEK, HOWARD. *Urban-Rural Integration in Regional Development: A Case Study of Sa. rashtra, India—1800–1960.* 1976. 156 pp.

172. COHEN, YEHOSHUA S. and BERRY, BRIAN J. L. *Spatial Components of Manufacturing Chang* 1975. 272 pp.

3. HAYES, CHARLES R. *The Dispersed City: The Case of Piedmont, North Carolina.* 1976. 169 pp.

4. CARGO, DOUGLAS B. *Solid Wastes: Factors Influencing Generation Rates.* 1977. 112 pp.

5. GILLARD, QUENTIN. *Incomes and Accessibility. Metropolitan Labor Force Participation, Commuting, and Income Differentials in the United States, 1960–1970.* 1977. 140 pp.

6. MORGAN, DAVID J. *Patterns of Population Distribution: A Residential Preference Model and Its Dynamic.* 1978. 216 pp.

7. STOKES, HOUSTON H.; JONES, DONALD W. and NEUBURGER, HUGH M. *Unemployment and Adjustment in the Labor Market: A Comparison between the Regional and National Responses.* 1975. 135 pp.

9. HARRIS, CHAUNCY D. *Bibliography of Geography. Part I. Introduction to General Aids.* 1976. 288 pp.

0. CARR, CLAUDIA J. *Pastoralism in Crisis. The Dasanetch and their Ethiopian Lands.* 1977. 339 pp.

1. GOODWIN, GARY C. *Cherokees in Transition: A Study of Changing Culture and Environment Prior to 1775.* 1977. 221 pp.

2. KNIGHT, DAVID B. *A Capital for Canada: Conflict and Compromise in the Nineteenth Century.* 1977. 359 pp.

3. HAIGH, MARTIN J. *The Evolution of Slopes on Artificial Landforms: Blaenavon, Gwent.* 1978. 311 pp.

4. FINK, L. DEE. *Listening to the Learner. An Exploratory Study of Personal Meaning in College Geography Courses.* 1977. 200 pp.

5. HELGREN, DAVID M. *Rivers of Diamonds: An Alluvial History of the Lower Vaal Basin.* 1979. 399 pp.

6. BUTZER, KARL W., editor. *Dimensions of Human Geography: Essays on Some Familiar and Neglected Themes.* 1978. 201 pp.

7. MITSUHASHI, SETSUKO. *Japanese Commodity Flows.* 1978. 185 pp.

8. CARIS, SUSAN L. *Community Attitudes toward Pollution.* 1978. 226 pp.

9. REES, PHILIP M. *Residential Patterns in American Cities, 1960.* 1979. 424 pp.

0. KANNE, EDWARD A. *Fresh Food for Nicosia.* 1979. 116 pp.

2. KIRCHNER, JOHN A. *Sugar and Seasonal Labor Migration: The Case of Tucumán, Argentina.* 1980. 158 pp.

3. HARRIS, CHAUNCY D. and FELLMANN, JEROME D. *International List of Geographical Serials, Third Edition, 1980.* 1980. 457 p.

4. HARRIS, CHAUNCY D. *Annotated World List of Selected Current Geographical Serials, Fourth, Edition. 1980.* 1980. 165 p.

5. LEUNG, CHI-KEUNG. *China: Railway Patterns and National Goals.* 1980. 235 p.

6. LEUNG, CHI-KEUNG and GINSBURG, NORTON S., eds. *China: Urbanization and National Development.* 1980. 280 p.

7. DAICHES, SOL. *People in Distress: A Geographical Perspective on Psychological Well-being.* 1981, 199 p.

8. JOHNSON, JOSEPH T. *Location and Trade Theory: Industrial Location, Comparative Advantage, and the Geographic Pattern of Production in the United States.* 1981. 107 p.

99-200. STEVENSON, ARTHUR J. *The New York-Newark Air Freight System.* 1982. 440 p.

01. LICATE, JACK A. *Creation of a Mexican Landscape: Territorial Organization and Settlement in the Eastern Puebla Basin, 1520–1605.* 1981. 143 p.

02. RUDZITIS, GUNDARS. *Residential Location Determinants of the Older Population.* 1982. 117 p.

03. LIANG, ERNEST P. *China: Railways and Agricultural Development, 1875–1935.* 1982. 186 p.

04. DAHMANN, DONALD C. *Locals and Cosmopolitans: Patterns of Spatial Mobility during the Transition from Youth to Early Adulthood.* 1982. 146 p.

05. FOOTE, KENNETH E. *Color in Public Spaces: Toward a Communication-Based Theory of the Urban Built Environment.* 1983. 153 p.

06. HARRIS, CHAUNCY D. *Bibliography of Geography. Part II: Regional. Vol. 1. The United States of America.* 1984. 178 p.

07-208. WHEATLEY, PAUL. *Nāgara and Commandery: Origins of the Southeast Asian Urban Traditions.* 1983. 473 p.

09. SAARINEN, THOMAS F.; SEAMON, DAVID; and SELL, JAMES L., eds. *Environmental Perception and Behavior: An Inventory and Prospect.* 1984. 263 p.

10. WESCOAT, JAMES L., JR. *Integrated Water Development: Water Use and Conservation Practice in Western Colorado.* 1984. 239 p.

11. DEMKO, GEORGE J., and FUCHS, ROLAND J., eds. *Geographical Studies on the Soviet Union: Essays in Honor of Chauncy D. Harris.* 1984. 294 p.

212. HOLMES, ROLAND C. *Irrigation in Southern Peru: The Chili Basin.* 1986. 191 p.

213. EDMONDS, RICHARD L. *Northern Frontiers of Qing China and Tokugawa Japan: A Comparative Stud of Frontier Policy.* 1985. 155 p.

214. FREEMAN, DONALD B., and NORCLIFFE, GLEN B. *Rural Enterprise in Kenya: Development and Spatia Organization of the Nonfarm Sector.* 1985. 180 p.

215. COHEN, YEHOSHUA S., and SHINAR, AMNON. *Neighborhoods and Friendship Networks: A Study o Three Residential Neighborhoods in Jerusalem.* 1985. 129 p.

217-218. CONZEN, MICHAEL P., ED. *World Patterns of Modern Urban Change: Essays in Honor of Chaunc D. Harris.* 1986.

219. KOMOGUCHI, YOSHIMI. *Agricultural Systems in the Tamil Nadu: A Case Study of Peruvalanallu Village.* 1986. 171 p.

220. GINSBURG, NORTON; OBORN, JAMES; and BLANK, GRANT. *Geographic Perspectives on the Wealth c Nations.* 1986. 131 p.